MODELING AND DESIGN TECHNIQUES FOR RF POWER AMPLIFIERS

THE WILEY BICENTENNIAL–KNOWLEDGE FOR GENERATIONS

Each generation has its unique needs and aspirations. When Charles Wiley first opened his small printing shop in lower Manhattan in 1807, it was a generation of boundless potential searching for an identity. And we were there, helping to define a new American literary tradition. Over half a century later, in the midst of the Second Industrial Revolution, it was a generation focused on building the future. Once again, we were there, supplying the critical scientific, technical, and engineering knowledge that helped frame the world. Throughout the 20th Century, and into the new millennium, nations began to reach out beyond their own borders and a new international community was born. Wiley was there, expanding its operations around the world to enable a global exchange of ideas, opinions, and know-how.

For 200 years, Wiley has been an integral part of each generation's journey, enabling the flow of information and understanding necessary to meet their needs and fulfill their aspirations. Today, bold new technologies are changing the way we live and learn. Wiley will be there, providing you the must-have knowledge you need to imagine new worlds, new possibilities, and new opportunities.

Generations come and go, but you can always count on Wiley to provide you the knowledge you need, when and where you need it!

WILLIAM J. PESCE
PRESIDENT AND CHIEF EXECUTIVE OFFICER

PETER BOOTH WILEY
CHAIRMAN OF THE BOARD

MODELING AND DESIGN TECHNIQUES FOR RF POWER AMPLIFIERS

Arvind Raghavan
Intel Corporation
Hudson, MA

Nuttapong Srirattana
RF Micro Devices
Greensboro, NC

Joy Laskar
Georgia Institute of Technology
Atlanta, GA

IEEE PRESS

WILEY-INTERSCIENCE
A John Wiley & Sons, Inc., Publication

Published by John Wiley & Sons, Inc., Hoboken, New Jersey.
Published simultaneously in Canada.

For general information on our other products and services or for technical support, please contact our Customer Care Department within the United States at (800) 762-2974, outside the United States at (317) 572-3993 or fax (317) 572-4002.

Wiley also publishes its books in a variety of electronic formats. Some content that appears in print may not be available in electronic format. For information about Wiley products, visit our web site at www.wiley.com.

Wiley Bicentennial Logo: Richard J. Pacifico

Library of Congress Cataloging-in-Publication Data:

Raghavan, Arvind.
Modeling and design techniques for RF power amplifiers / by Arvind Raghavan, Nuttapong Srirattana, Joy Laskar.
p. cm.
Includes bibliographical references and index.
ISBN 978-0-471-71746-1 (cloth)
1. Power amplifiers—Design and construction. 2. Amplifiers, Radio frequency.
I. Srirattana, Nuttapong. II. Laskar, Joy. III. Title.
TK7871.58.P6R34 2007
621.384'12—dc22 2007027547

Printed in the United States of America.

10 9 8 7 6 5 4 3 2 1

CONTENTS

PREFACE

Wireless communication is ubiquitous in today's world. The advancement of wireless communication technology in the quest for better performance at lower cost has resulted in increasingly stringent demands on the integrated circuits (ICs) that constitute the building blocks of wireless systems. The radiofrequency (RF) power amplifier, an important component of any wireless transmitter, is often the villain of the piece, since it is a limiting factor in achieving better performance and reliability, and lowering cost. Thus, RF power amplifier design is a topic of immense interest and import in wireless communications.

This book discusses different aspects of RF power amplifier design. RF power amplifier design is multidisciplinary, and requires the designer to be cognizant of the several factors that have an impact on the performance of the power amplifier IC: the choice of semiconductor technology, accuracy of the transistor device models, packaging and thermal management, and circuit and architectural design techniques. Most textbooks on RF power amplifiers discuss only circuit design and architecture. The scope of this book extends beyond that, and a discussion of device models and device physics concepts, knowledge of which is of great value to the power amplifier designer, is included. RF power amplifier design is also an area of active research. An objective of this book is to acquaint the reader with the latest advances in power amplifier design. For each topic dealt with in this book, a discussion of the fundamental concepts is followed by a description of recent developments.

For a long time, power amplifier design, especially for mobile applications, remained the domain of III–V semiconductor technologies. This has changed recently, with SiGe HBT technology becoming an option for building RF power amplifiers. Power amplifier design in standard silicon-based complementary metal oxide semiconductor (SiCMOS) technology is the subject of much current research. Silicon-based power amplifiers create interesting possibilities for higher integration and lower cost. However, power amplifier design in silicon, especially in CMOS, poses considerable challenges. This book treats the important topic of power amplifier design in silicon, and includes a discussion of state-of-the-art design techniques in this area.

Another trend that has emerged in recent wireless communication standards is the use of modulation schemes with significant amplitude modulation, to achieve higher spectral efficiency. This has the effect of increasing the power amplifier linearity requirement, which conflicts with the desire for higher efficiency, an important criterion in mobile handsets, since it directly affects talk time. Thus, architectural techniques to enhance the efficiency of linear power amplifiers, a topic discussed in this text, have become very important in modern RF power amplifier design.

Accurate transistor device modeling is important in today's competitive IC design world, with increasing emphasis on fast design cycles and first-pass design success. This is particularly true for RF power amplifier design, since the power amplifier usually operates at the limit of the capability of the semiconductor device technology. Traditional device models typically prove inadequate to the task of accurately predicting power amplifier performance. New device models and enhancements to existing models have been proposed in recent years, to overcome the deficiencies of older models. An understanding of the formulation, features, and shortcomings of various device models, and the underlying device physics, enables the power amplifier designer to make more informed decisions about the choice of device technology and appropriate model to be used for a particular circuit design. It also provides the designer with an understanding of the expected accuracy of the predicted circuit performance. This book describes various bipolar transistor and metal oxide semiconductor field effect transistor (MOSFET) device models, and discusses advanced modeling techniques pertaining to RF IC design.

The book is organized as follows. Chapter 1 introduces RF power amplifier design concepts. An overview of considerations related to various topics of interest to the power amplifier designer, such as semiconductor technologies, simulation tools, and modulation schemes, to name a few, is provided. Chapter 2 discusses bipolar transistor and MOSFET device physics and device models. Chapter 3 describes an empirical modeling technique applicable to a wide variety of bipolar transistors. The goal of this technique is to simplify model parameter extraction while retaining sufficient accuracy for RF circuit design. Chapter 4 discusses enhancements to a standard MOSFET model, BSIM3v3, to model the RF characteristics more accurately. This includes a scalable model with substrate network. Chapter 5 covers power amplifier IC design concepts. This chapter includes discussions of power amplifier design methodology, classes of operation, and power amplifier performance metrics. Chapter 6 is concerned with silicon-based power amplifier design. Design techniques are illustrated using practical design examples. A SiGe heterojunction bipolar (HBT)-based high-efficiency power amplifier module is presented. This design demonstrates the integration of power amplifier IC and novel board-level design techniques using a high-performance LTCC substrate, to improve the overall performance of the power amplifier. This chapter also discusses RF power amplifier design in CMOS. A novel device periphery adjustment technique to improve the efficiency of power amplifiers at low output power levels is described. Chapter 7 deals with efficiency enhancement architectures, the focus being on the multistage Doherty amplifier technique.

This text compiles discussions of some of the most important topics related to RF power amplifier design in one concise volume. It is envisioned to serve the purposes of a wide range of readers, from beginning designers to advanced researchers. The discussion of the fundamental concepts of a range of topics pertaining to power amplifier design makes this book suitable for use as a textbook or reference for graduate or advanced undergraduate courses in RF IC design. The inclusion of recent developments serves to introduce beginners to more advanced topics, and is also particularly useful to researchers in this area. The description of state-of-the-art techniques, without sacrificing breadth of coverage of topics, is intended to make this book a valuable and handy reference to practicing engineers.

This book was made possible by assistance from many individuals and organizations. The material covered in this book evolved over several years of research and teaching at the Georgia Institute of Technology. We would like to acknowledge our colleagues, past and present, at the Georgia Institute of Technology, particularly Prof. Phillip Allen, Prof. John Cressler, Dr. Deukhyoun Heo, Dr. Kyutae Lim, Dr. Stephane Pinel, Dr. Moonkyun Maeng, Dr. Albert Sutono, Dr. Sebastien Nuttinck, Dr. Ramana Murty, Dr. Emery Chen, Dr. Chang-Ho Lee, Dr. Sudipto Chakraborty, Dr. Bhaskar Banerjee, Sunitha Venkataraman, and Anand Raghavan. Special thanks are due to Patrick O'Farrell, Andy McLean, and Arlo Aude of National Semiconductor. We are grateful to the staff at John Wiley & Sons, Inc., especially Whitney Lesch, for their assistance. Most importantly, we thank our respective families for their patience and support, especially Vasantha Raghavan, Bashyam Raghavan, Noppawan Srirattana, Chongrak Srirattana, Nuttapan Srirattana, Thidarat Tosukhowong, Devi Laskar, Anjini Laskar, Ellora Laskar, and Devrani Laskar.

Arvind Raghavan
Nuttapong Srirattana
Joy Laskar

1

INTRODUCTION

The radiofrequency (RF) power amplifier is one of the most important components of a wireless communication system. It plays a significant part in determining the overall performance, cost, and reliability of the wireless system. In fact, the increasing use of modulation schemes with amplitude modulation of the carrier signal, to achieve higher spectral efficiency, is only expected to enhance this significance. The RF power amplifier is also one of the most challenging components of the wireless system front end to design, analyze, and model for, as it generally operates at the limits of the capability of the semiconductor device technology. This situation is exacerbated by the current trend toward migration of the radio front-end to silicon-based technologies, which are considerably less suitable for RF power amplifier design than are the traditional III–V-based technologies. RF power amplifier design encompasses, and is impacted by, several areas: semiconductor technology, which provides the active and passive devices used in power amplifier design and determines the performance and reliability of the power amplifier to a large extent; transistor device modeling, which makes it possible to design and predict the behavior of the power amplifier; RF measurement and characterization techniques; integrated circuit (IC) design; architectural techniques, which help improve the performance of the power amplifier; behavioral "blackbox" modeling, which enables analysis of the impact of the characteristics of the power amplifier on the performance of the wireless system; and IC packaging and thermal management technology.

1.1 SEMICONDUCTOR TECHNOLOGY AND RF POWER AMPLIFIER DESIGN

A wide variety of semiconductor technologies have been used for RF power amplifier design: gallium arsenide (GaAs)-based heterojunction bipolar transistors (HBTs), metal semiconductor field effect transistors (MESFETs), high-electron-mobility transistors (HEMTs), and pseudomorphic HEMTs (or pHEMTs), silicon bipolar junction transistors (Si BJTs), silicon metal oxide semiconductor field effect transistors (Si MOSFETs), silicon laterally diffused metal oxide semiconductor (LDMOS) FETs, silicon germanium (SiGe) HBTs, and gallium nitride (GaN)-based HEMTs. The choice of technology is dictated by the suitability of the properties of the semiconductor device for a particular application. For instance, Si LDMOS technology is widely used to build power amplifiers for base-station transmitters in cellular networks. This is because it is a mature device technology; offers a good combination of gain, linearity, reliability, and cost; and is capable of delivering tens or hundreds of watts of output power from tens of volts of power supply. It is worth emphasizing the role of cost in the choice of semiconductor technology. For example, in the case of base-station power amplifiers, high-power GaAs pHEMTs and MESFETs have been shown to exhibit superior performance compared to Si LDMOS FETs. However, the higher cost of these devices has meant that LDMOS technology continues to dominate this application, especially at frequencies below 2 GHz.

The quest for new materials and technologies, improvement of the characteristics of existing semiconductor devices, and better processing techniques, to reduce the cost of a technology, is a constant pursuit, and the subject of active research and development. Good examples of emerging technology are the wide-bandgap semiconductors, silicon carbide (SiC) and gallium nitride (GaN). Their excellent material properties, such as high breakdown voltage (their breakdown electric field is over 5 times higher than that of Si or GaAs), high saturated electron drift velocity, and high thermal conductivity (especially that of SiC), have generated an enormous amount of interest in their potential for high-power applications.

GaAs HBTs and GaAs MESFETs have been the mainstays of RF power amplifier design for mobile handsets for many years. In handset applications, the maximum power is of the order of 1 W. Device area is a critical concern, since compactness is an essential feature. GaAs HBTs in particular, enjoy an advantage in this respect, because of their high-power density, which originates from their high-current-handling capability. The area of the power transistor for a given output power requirement is, therefore, typically smaller than that of most other devices for the GaAs HBT (even other GaAs devices such as GaAs FETs), despite the GaAs HBT requiring additional circuitry such as ballast resistors. GaAs HBTs posses other advantages over GaAs MESFETs, such as requiring only a single positive supply voltage, low leakage current requiring no extra DC switch to turn off the power supply in standby mode, and breakdown voltage being independent of input voltage. As a result, they have come to dominate the handset power amplifier market. However, GaAs HBTs suffer from thermal issues, such as self-heating and current gain collapse, which have to be carefully managed. Self-heating can lead to a significant power difference between continuous and pulsed modes of operation in a GaAs HBT, unlike a GaAs MESFET.

GaAs HBTs usually use the AlGaAs/GaAs configuration. Here, a bandgap difference is introduced by using two different materials for the emitter and the base (i.e., AlGaAs and GaAs, respectively). The idea of improving device performance by introducing a bandgap difference was conceived by Shockley in 1948, but fabrication technology was not sufficiently developed to accomplish this until the development of molecular beam epitaxy (MBE) in the mid-1970s [1]. This bandgap difference increases the current gain of the device, which can be achieved while simultaneously increasing base doping to reduce its resistance. The reduced base resistance enables a higher frequency of operation. The current flow is vertical, so surface defects do not have a big effect on performance. More recently, HBTs based on the structure InGaP/GaAs have found use in RF power amplifier design. One advantage of InGaP/GaAs HBTs is that they do not suffer from early current gain collapse as do the AlGaAs/GaAs HBTs. The variation in current gain over operating temperature range for an InGaP/GaAs HBT is much smaller than that for an AlGaAs/GaAs HBT.

In recent times, there has been a great interest in building RF power amplifiers in silicon-based technologies. A silicon implementation of the RF power amplifier would go a long way toward achieving the holy grail of "system on a chip," where the whole radio system would be integrated on a single die, enabling large cost savings. SiGe HBT technology has been the pioneering technology in the effort to build the RF power amplifier in silicon, with commercial power amplifier products now available. Building power amplifiers in Si CMOS, on the other hand, is more challenging, and considerable effort is currently being expended in this area. The considerations and challenges in designing RF power amplifiers in silicon, and advanced techniques to overcome these challenges, are discussed in Chapter 6.

1.2 DEVICE MODELING

Transistor device models are indispensable in modern IC design, most of which is computer-aided. In the early days of the use of transistors in electronic circuits, the design of these circuits relied heavily on empirical methods. The designer actually built the conceived circuit on a circuit board with discrete elements, and tested its electrical performance. The elements were consequently changed in value and configuration, until the desired specifications were met. However, this "breadboarding" methodology became infeasible with the advent of the integrated circuit (IC), because the complexity of circuits that could be designed increased enormously, and because parasitic effects and coupling between various devices on the same substrate became significant. As a result, it became essential to develop "equivalent circuit" models for the quantitative terminal description of the transistor (and other electrical elements), which could be used to "simulate" the performance of the proposed circuit configuration. The emergence of device models enhanced flexibility in IC design, allowed for components that were difficult to breadboard, and shortened the circuit design-cycle. Transistor models used for circuit simulation are usually called "compact" models, as distinct from the more elaborate numerical device simulation models based on fundamental device physics, which are used for device design. RF power amplifier design, like any other circuit block in the radio front end, relies on accurate transistor models to predict the performance of the power amplifier, and hence enable it to be designed efficiently. Over the years, several different models have been developed for both bipolar and FET devices. However, many of the traditionally used models are not very suitable for RF power amplifier design. This is because many effects that are important from in terms of power amplifier design, such as self-heating effects in bipolar transistors, especially those in GaAs-based processes, distributed effects in the gate and substrate of Si MOSFETs, and scaling of transistor model parameters with power cell size, are not adequately modeled. Chapters 2, 3, and 4 provide a description of various device models for bipolar devices and MOSFETs, and discuss advanced modeling techniques.

1.3 POWER AMPLIFIER IC DESIGN

The design of RF power amplifiers is often regarded as being as much an art as a science. Traditionally, power amplifiers are designed as much by experimental iteration and tweaking, as by computer-aided design (CAD) techniques and simulation. The large-signal behavior of the RF power transistor has often been a source of controversy and debate. The optimum output power match impedance was regarded as something that could be measured only experimentally, by "load-pull" measurements, an indispensable, albeit expensive tool, in power amplifier design. However, the constant power contours obtained in load-pull measurements are not as mysterious as they are made out to be, and it is possible to reconcile them with simple load-line principles [2]. Another reason for the confusion surrounding power amplifier design is the multitude of classes of power amplifier operation. The behavior of the RF power transistor can be quite complex; it can act either as a high-resistant current source, or as a low-resistance switch, or, in some amplifiers, as a high-resistance current source during one part of the cycle, and low-resistance switch during another part of the cycle (mixed-mode operation). Further, the same circuit topology can operate in different modes, depending on how the transistor is biased and on specific inductance–resistance–capacitance (*L–R–C*) values in the load network. In fact, the power amplifier often operates in modes unbeknownst to, and unintended by, the designer, because of insufficient analysis of the design, occasionally even resulting in better performance than anticipated. Indeed, the understanding of the nuances of power amplifier operation is far from complete, and is still a subject of research. Chapter 5 provides an introduction to the various concepts and considerations in RF power amplifier IC design. A discussion of the main power amplifier classes is included, as is a summary of various power amplifier performance metrics. Chapter 6 discusses silicon-based power amplifier design, a topic of great interest today. Chapter 7 focuses on efficiency enhancement techniques for linear power amplifier design. This is also a topic of much importance, in view of the increasing use of nonconstant-envelope modulation schemes in modern wireless communication systems.

1.4 POWER AMPLIFIER LINEARITY

A fundamental concern when transmitting an RF signal is that it must not interfere with transceivers operating in adjacent channels. However, nonlinearity in the wireless transmitter (especially the power amplifier) causes distortion of the signal, which results in the bandwidth of the signal spreading out into adjacent channels. In general, two types of distortion occur when a signal passes through a nonlinear circuit. One is harmonic distortion, which causes the signal to be replicated at harmonic frequencies of the carrier, and the other is intermodulation distortion, which adds a "skirt" to the signal, and causes its bandwidth to spread. Harmonic distortion is the easier of the two to deal with, as it can be removed using a bandpass filter. However, the intermodulation distortion products overlap (and add skirts to) the original signal bandwidth, and cannot be removed by filtering. Thus, some amount of the transmitted power will leak into adjacent channels. This spreading of the signal bandwidth due to intermodulation distortion is called *spectral regrowth*. To allow the operation of many different communication channels in the available wireless spectrum, there are specified limits to the allowed level of spectral regrowth or adjacent-channel interference. Thus, the linearity of a power amplifier is a critical consideration in its design. For a given power amplifier design, the degradation it produces in the spectral regrowth depends on the nature of the envelope of the modulated RF signal. A constant envelope results in lesser degradation than a variable envelope. In other words, if the modulated RF signal has a nonconstant envelope, it will require a highly linear power amplifier to satisfy the adjacent-channel interference requirements. The linearity of the power amplifier, in addition to affecting signals in neighboring channels, also affects the achievable bit error rate of the wireless communication system. Linearity metrics for RF power amplifiers are discussed in greater detail in Chapter 5.

1.5 MODULATION SCHEMES

The constant or variable nature of the envelope of the RF carrier is the consequence of the modulation scheme used in the wireless communication system. Modulation is the method by which the information desired to be transmitted is encoded onto the RF carrier. In analog modulation, some parameter of the transmitted signal (the carrier) is varied as a linear function of the amplitude of the original audio or video signal (the modulating signal) to be transmitted. The parameters of the carrier that can be modulated are its amplitude, phase, and frequency. Frequency and phase modulation are less sensitive to nonlinearities in the amplitude response of RF circuits, noise, and time-varying fading in the channel, than amplitude modulation, but this improvement in performance is at the expense of increased transmission bandwidth. The purpose of digital modulation is to convert an information-bearing discrete-time symbol sequence into a continuous-time waveform. In digital modulation, the signal to be transmitted is typically binary data, which may, interestingly, be a quantized and digitized analog signal. A simple form of digital modulation is on-off keying (OOK), in which a carrier is turned on and off, depending on the value of each bit. The generalized version of this is amplitude-shift keying (ASK), where the amplitude of the carrier is varied in discrete steps, with each level representing one or more bits. ASK demonstrates poor performance, as it is heavily affected by noise and interference. Frequency-shift keying (FSK) uses discrete frequencies as the symbols. For example, to transmit binary data, two different frequencies slightly offset from the carrier frequency are usually used. Phase-shift keying (PSK) uses a carrier of constant nominal frequency, with different phase shifts in the phase of the carrier signal relative to a reference phase used to represent different symbols. For example, in binary phase-shift keying (BPSK), a phase shift of π may be introduced when the transmitted symbol changes from 0 to 1, and the same would happen when it changes from 1 to 0. BPSK demonstrates better performance than ASK and FSK, and filtering can be employed to limit spectral spreading. However, the transmitter and receiver are also more complex. Quadrature phase shift keying (QPSK) is an improvement over basic BPSK. The phasor representation of BPSK and QPSK, shown in Figure 1.1, illustrates the difference between the two. In QPSK, a higher bit rate is achieved for the same bandwidth by coding two bits into one phase shift. QPSK is effectively two independent BPSK systems (called in-phase I, and quadrature Q), and therefore exhibits the same performance at twice the bandwidth efficiency. In PSK systems, instead of using phase shifts relative to a reference signal of the same frequency, a phase shift relative to the previously transmitted signal can used; this is differential phase shift keying. Such a scheme is preferable in mobile systems, since the phase of the received signal changes rapidly and it is difficult to maintain a constant phase reference. An important issue in wireless communication is the spectral occupancy, or bandwidth, of the transmitted signal. It is desirable to limit the bandwidth of the signal, to improve spectral efficiency, and reduce interference with adjacent channels. Filtering is used for this purpose; root raised-cosine filters are popular because they offer an approximation to the minimum required bandwidth (the Nyquist bandwidth). For instance, in QPSK, raised-cosine filters are used to achieve good out-of-band suppression. Filtered QPSK exhibits a nonconstant envelope, and therefore a linear power amplifier is required in the transmitter. Conventional QPSK is also not very spectrally efficient, due to the instantaneous π phase shift. In offset QPSK, the I and Q channels are staggered, and phase transitions are therefore limited to $\pi/2$. In $\pi/4$-QPSK, the set of constellation points are toggled each symbol, so transitions through zero cannot occur. These schemes produce smaller envelope variations compared to conventional QPSK. QPSK-based modulation formats are used in NADC (North American Digital Cellular), Japanese PHS (personal handy-phone system), and code-division multiple-access (CDMA) systems. Minimum-shift keying (MSK) is a form of continuous-time FSK, where the phase is changed between symbols so as to provide a constant envelope. In MSK, phase ramps up by $\pi/2$ for, say, a binary 1, and down by $\pi/2$ for 0. Adding a Gaussian lowpass filter to the MSK scheme results in the so-called Gaussian MSK (GMSK), which is a popular constant-envelope modulation format, and is used in GSM (global system for mobile communications) systems. However it is spectrally less efficient than filtered QPSK modulation. Also, if the bandwidth–bit period product (or BT) is too low, significant intersymbol interference (ISI) is created.

Amplitude and phase shift keying can be combined to transmit several bits per symbol. Such modulation schemes require linear power amplifiers. Higher linearity in power amplifiers is achieved at the expense of efficiency. Thus, there is a fundamental tradeoff between spectral efficiency and power efficiency in wireless transmission. Quadrature amplitude modulation (QAM) is example of a modulation format which uses symbols that vary in both amplitude and phase. An example of this type of modulation, 16-QAM, is shown in Figure 1.1. Such multilevel (or M-ary) modulation formats are, in general, more bandwidth-efficient, but are also more susceptible to noise.

Orthogonal frequency-division multiplexing (OFDM) is a multicarrier modulation technique, where the source symbols are transmitted in parallel using many orthogonal subcarriers. IEEE 802.11a/g wireless LAN (local area network) systems use OFDM. Multicarrier modulation exhibits good ISI mitigation. Also, frequency-selective fading may influence only some subcarriers, and not the whole signal. However, multicarrier modulation techniques result in a more stringent linearity requirement on the power amplifier, because they exhibit a large peak-to-average power ratio (PAPR). Ultra-wideband (UWB) is another, relatively new technique, used in wireless transmission. Pulses with a high bandwidth (1 GHz or more) are used for transmission in a UWB system. Pulse position modulation, where the value of a transmitted symbol is given by the precise timing of the pulse, is a modulation format suitable for UWB systems.

Spread-spectrum techniques are a category of techniques used to combat narrowband interference, or cochannel interference, between modulated signals, by spreading each signal over a wider bandwidth using a code that is known to both the transmitter and the receiver. In frequency-hopping spread spectrum (FHSS) systems, the total available bandwidth is split into many channels of smaller bandwidth. The transmitter transmits on one of these channels for a certain period of time, and then hops to another channel. Thus FHSS implements both frequency- and time-division multiplexing. The hopping may be slow, where the transmitter uses one frequency for several bit periods, or fast, where the transmitter may change the frequency several times during a bit period. Fast hopping systems offer greater tolerance to narrowband interference and frequency-selective fading, at the cost of increased complexity. An example of an FHSS system is Bluetooth. In direct-sequence spread-spectrum (DSSS) systems, the bit sequence is multiplied by a binary pseudonoise (PN) sequence. The PN sequence rate that is higher than the bit rate, and each data bit is broken into several chips, where each chip is the product of the data bit and a digit of the PN sequence. This process spreads the signal power over a wider bandwidth. The bit stream is recovered by correlating the received signal with an identical PN sequence. If the transmitter and receiver are perfectly synchronized and the signal is not too distorted by noise or multipath fading, it is easy to recover the transmitted data using a correlator. In a real-world situation, where effects such as multipath fading exist, recovery is complex. A rake receiver, which uses n correlators for the n strongest paths, is used in such a case. Each correlator is synchronized to the transmitter plus the delay on that specific path. As soon as the receiver detects a new path that is stronger, it assigns this path to the correlator with currently the weakest path. DSSS can be used as a form of multiple access, by assigning different PN sequences that have low cross-correlation, to several users. The users can then share the same frequency spectrum because only the desired transmitter will have a high correlation with the PN code used at each receiver.

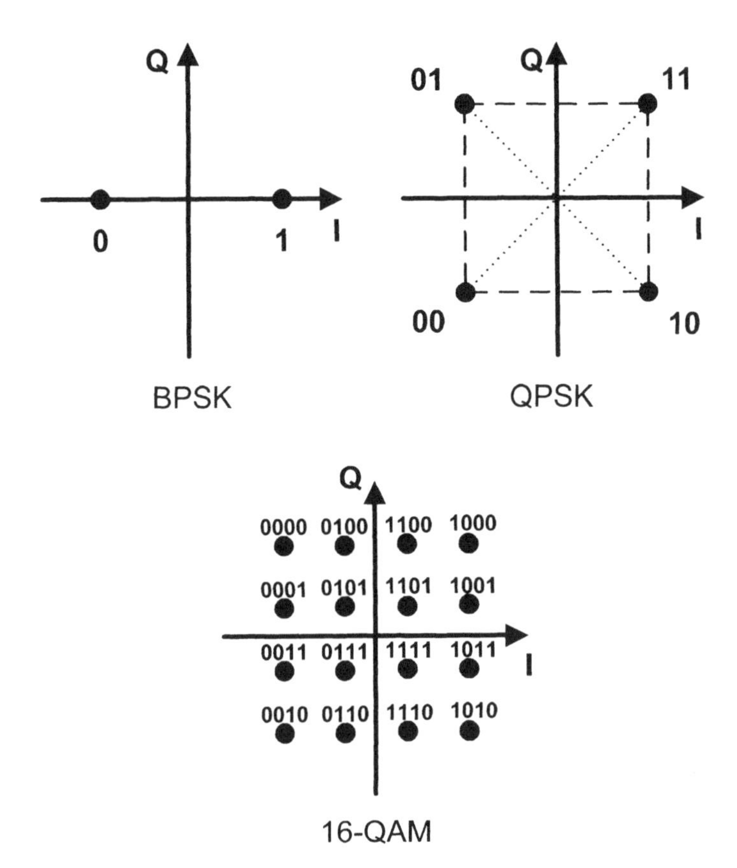

Figure 1.1. Phasor representation, or constellation diagrams, for BPSK, QPSK, and 16-QAM modulation schemes.

1.6 CIRCUIT SIMULATION

Simulation tools play a vital role in the design of the RF power amplifier IC. They enable computer-aided design by utilizing transistor and passive device models to solve for the various voltages and currents in the electric circuit being designed, making it possible to predict its characteristics. SPICE (simulation program with integrated circuit emphasis) is an example of a circuit simulator. In fact, it has (in its various versions) for long been the most popular circuit simulation tool. However, RF circuits present a unique problem for traditional transient analysis using SPICE. This is because, in wireless communication, the signal comprises a high-frequency carrier, modulated by a low-frequency modulation signal. The high-frequency carrier requires a small time step in such a transient analysis, while the low-frequency modulation necessitates a long simulation interval. Thus, modulated RF signals represent a worst-case scenario for the efficiency of SPICE transient analysis. Techniques more suitable for RF IC design have been developed, and in general, these can be classified into two groups: harmonic balance-based methods and shooting methods [3]. Harmonic balance and shooting methods have been incorporated in commercially available and widely used RF circuit simulators. These techniques compute the periodic steady-state solution for a circuit, which is the steady-state response of a circuit driven with periodic waveforms. The steady-state response of a circuit driven by one or more large periodic signals can also be computed using quasiperiodic analysis techniques. The solution of such an analysis is used as a periodic or quasiperiodic operating point for further simulations. Harmonic balance techniques formulate the circuit equations in the frequency domain. Shooting methods are designed to solve boundary-value problems, using iterative techniques in addition to transient analysis. Newton methods figure prominently in shooting-method-based simulators, and such an algorithm is called the *shooting-Newton algorithm*. Harmonic balance is very efficient for circuits that are not very nonlinear, that is, if the distortion levels are low. It is also very accurate if the stimuli are sinusoidal. However, harmonic balance poses problems with circuits that are highly nonlinear, because, a large number of frequencies are needed to accurately represent the signals, which considerably slows down the analysis. The harmonic balance method also encounters convergence issues with strongly nonlinear circuits. However, techniques have been developed to improve the convergence of harmonic balance under such conditions. This involves initially reducing the amplitude of the input signal to achieve convergence, and subsequently increasing the amplitude in steps, using the result computed at one step as the starting point for the next one. Such techniques, called *continuation* or *homotopy methods*, improve convergence of harmonic balance, but at the expense of speed. Shooting methods, on the other hand, are intrinsically better at handling strongly nonlinear circuits, since they use nonuniform time steps. Also, since they use Newton's method to iteratively arrive at a solution, their convergence under nonlinear conditions is very good. It is worth noting that many wireless communication applications use nonconstant-envelope modulation and require a high level of linearity in the power amplifier. In such a situation, harmonic balance is quite a suitable and efficient tool for circuit simulation. Harmonic balance is also very adept at handling transmission lines, and table-based interpolated *S*-parameter models. Shooting methods, on the other hand struggle to handle distributed components, such as transmission lines. This is why harmonic-balance-based simulators have been more popular in microwave design, in which such components are often used. System-level simulation tools, which allow cosimulation with embedded circuit components, have also successfully used harmonic balance for the analysis of these circuit blocks.

Spectral regrowth is very difficult to predict with traditional SPICE transient analysis. This is because the high carrier frequency, and the fact that a large number of bit transmissions (hundreds or a few thousand) must be simulated to obtain a reasonably representative spectrum, makes the use of traditional transient analysis impractical. The stimulus signal, generated by the digital modulation format in use in the wireless communication system, has to be accurately represented. Transient envelope analysis may be used to simulate modulated carrier systems. In such an analysis, a series of linked large-signal pseudoperiodic analyses, which are periodic analyses that have been modified to account for slow variations in the envelope as a result of modulation, are performed [3]. The pseudoperiodic analyses are performed often enough to adequately capture the changes in the envelope. Harmonic-balance based envelope simulation techniques, where the amplitude of the frequency components can vary as a function of time, have also been developed to simulate modulated RF signals and analyze spectral regrowth. However, because of the large number of simulation points usually required to simulate digitally modulated signals accurately, it is computationally quite expensive to use transient envelope techniques based on harmonic balance or shooting methods. An alternative is to use a behavioral model of the power amplifier, which is usually extracted from AM-AM and AM-PM measurements. Spectral regrowth can then be quickly and efficiently computed using this behavioral model and an appropriate mathematical characterization of the stimulus signal.

1.7 LOAD-PULL MEASUREMENTS

Load-pull measurements are widely used in RF power amplifier design and characterization. In a load-pull measurement, a whole range of output impedances are presented to the power transistor or power amplifier, and the characteristics of the device, like output power, power-added efficiency, intermodulation distortion, or adjacent-channel rejection, may be measured for each output impedance condition. The results are usually plotted on a Smith chart, to generate the so-called load-pull contours. Similar to load-pull, the source impedance may also be varied, and such a measurement is called *source-pull*. The output or input impedances are varied using tuners, which may be passive or active. A typical passive tuner system is shown in Figure 1.2. The two tuners may be used to simultaneously tune the source and load impedances at the fundamental frequency. Load-pull measurements are often used to optimize the match impedance, to extract the best possible performance from the power amplifier. Figure 1.3 shows typical load-pull contours for output power and power-added efficiency (PAE). Each of the contours represents the set of terminations corresponding to a particular output power level or value of PAE.

The termination impedance for which the output power (or PAE) is maximum is the center of the concentric set of contours. The optimum output impedance for maximum output power, for a power amplifier, is generally close to the periphery of the Smith chart. This means that the magnitude of the optimum impedance is small, which is expected since the output stages are large transistors carrying large current. This leads to practical difficulties in performing load-pull measurements, since losses in the RF signal path in the measurement setup reduces the effective area of the Smith chart that can be covered by the system. Prematching tuners may be used to shift the impedance of the device under test (DUT) to a range that can be easily achieved with the tuner setup, if the optimum impedance lies outside its range. The shape of the load-pull contour can be explained to consist of two arcs of constant power on the Smith chart [2]. For a given output power level, the set of output impedances corresponding to this power level consist of arcs of constant resistance (say R_{opt}) and constant conductance ($G_{opt} = 1/R_{opt}$) on the Smith chart, which intersect to give the load-pull contour.

In the system shown in Figure 1.2, the impedances at the harmonics of the fundamental frequency are not controlled. Harmonic terminations can be important in optimizing the performance of an RF power amplifier. Harmonic load- and source-pull measurements may be performed to characterize the effect of, and optimize, harmonic impedances. In a measurement where the second-harmonic impedances are being investigated, diplexers are used at the input and output of the DUT, to separate the signal paths at the fundamental and second-harmonic frequencies. Separate tuners are connected to each diplexer output, so that the terminations at the two frequencies can be tuned independently. Similarly, triplexers are used in a measurement involving third-harmonic terminations in addition to the fundamental and second-harmonic.

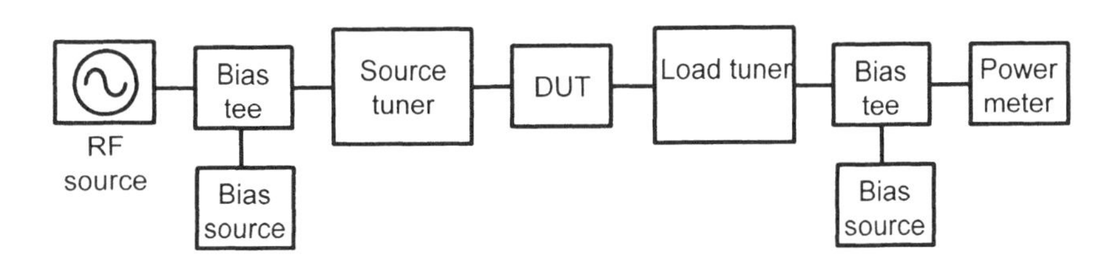

Figure 1.2. Load-pull system using passive tuners.

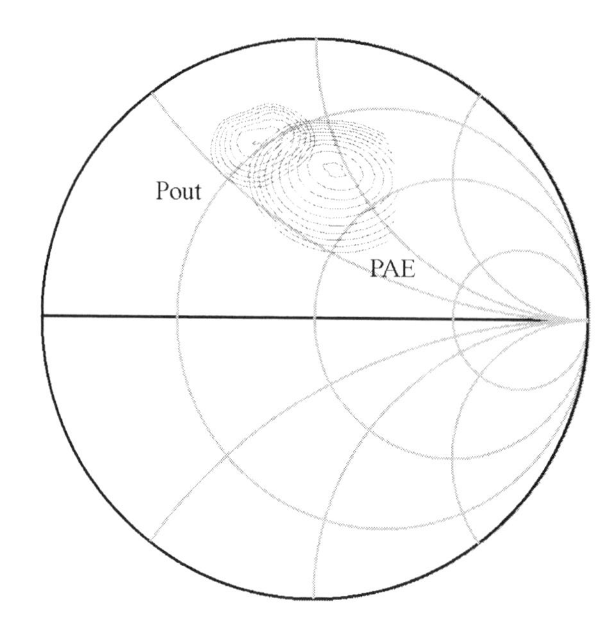

Figure 1.3. Typical load-pull contours for output power and PAE.

Active load-pull systems use active tuners, and are generally categorized into two types: the two-signal-paths type, and the feedback type [4]. In the two-signal paths type active load-pull system, shown in Figure 1.4a, a power divider splits the source RF signal into two parts. One drives the input port of the DUT, while the other is properly amplified, phase-shifted and injected into its output port. In the feedback type (see Figure 1.4b), a portion of the output signal of the DUT is drawn using a direction coupler and is fed back to the DUT output after proper amplification and phase shift. A high-selectivity filter has to be introduced in the loop, as shown in Figure 1.4b, to avoid oscillations, which this type of load-pull system is susceptible to, due to the broadband nature of the loop components. The two-signal-paths system is not susceptible to oscillations because of the high isolation of the power amplifiers. However, it suffers from the drawback that the load impedance also depends on the input power level and the DUT characteristics. Hence, a complicated sequence of attenuator and phase shifter adjustments is needed at each power level to keep the load impedance constant, during a power-sweep measurement.

Load/source-pull measurements find widespread application in designing and optimizing RF power amplifiers, and in characterizing, developing, and validating large-signal transistor models for use in power amplifier simulations.

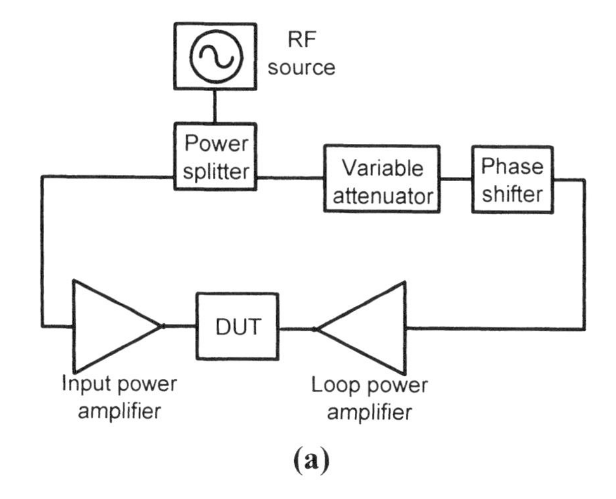

(a)

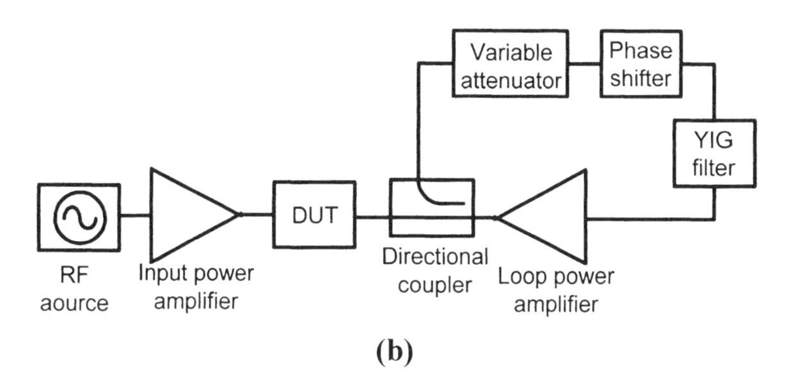

(b)

Figure 1.4. Active load-pull system configurations: (a) the two-signal-paths system; and (b) the feedback loop system.

REFERENCES

1. H. Kroemer, Heterojunction devices, *IEEE Trans. Electron Devices*, **39**(11): 2635–2636 (Nov. 1992).
2. S. Cripps, A method for prediction of load-pull power contours in GaAs MESFETs, *IEEE MTT-S Int. Microwave Symp. Digest*, 1983, pp. 221–223.
3. K. S. Kundert, Introduction to RF simulation and its applications, *IEEE J. Solid-State Circuits*, **34**(9): 1298–1319 (Sept. 1999).
4. J. E. Müller and B. Gyselinckx, Comparison of active versus passive on-wafer load-pull characterization of microwave and mm-wave power devices, *IEEE MTT-S Int. Microwave Symp. Digest*, 1994, pp. 1077–1080.

2

DEVICE MODELING FOR CAD

2.1 INTRODUCTION

The advent of SPICE (simulation program with integrated circuit emphasis) heralded the beginning of the era of computer-aided IC design. Electrical transistor models are indispensable for the simulation and design of integrated circuits. There are different approaches to defining a transistor device model, namely, the physical, the empirical, or a hybrid of the two. Models based on device physics are directly derived from the physical equations that describe the device. They consist of a set of analytical equations or functions, preferably in explicit form. The analytical nature of the functions is important, because discontinuities in the functions and their derivatives may cause numerical instabilities, and inaccuracies in the prediction of harmonics and distortion. One problem with physics-based models is that analytical expressions derived from device physics are typically valid only in a certain range of bias conditions, and outside that range other expressions must be used. Smoothing procedures utilizing curve fitting are usually applied to ensure continuous transition from one region to another. A benefit of physics-based models is that the parameters have physical significance, which can be used as a check on the correctness of parameter extraction, and is helpful in circuit design. Also, they have the ability to forecast the performance of new devices within their own class. Furthermore, geometric scaling rules can be applied with a degree of confidence. On the other hand, developing a physics-based model requires considerable time and effort, and this has to be repeated with every significant modification in the device structure.

Another class of models is table-based numerical models, which are based on the storage of measured data of the relevant electrical characteristics of a device. These models are quite accurate within the range of conditions at which the device was characterized, but require exhaustive characterization effort, and have no forecasting or scaling ability.

The other class of device models is the empirical model, in which the model-defining equations are analytical expressions, the parameters of which are derived by curve fitting to suitable measured data. These models require less exhaustive characterization effort compared to table-based models, and can be developed considerably faster than purely physical models. In practice, most physical models feature a certain degree of empiricism. In addition, the inability to arrive at closed-form analytical solutions for certain parameters in some devices necessitates the use of an empirical approach.

The choice of the approach to be adopted in modeling a device depends on the complexity of the physics of a given device, as well as the application for which the model is intended. An important issue is the tradeoff between complexity and accuracy. A complex model with detailed modeling of various physical effects in the device, some of which may not be important for the application in view, unnecessarily complicates the model extraction process, places a higher demand on the computational effort, and increases development time. A judicious choice of complexity of the model is an important factor in making the design cycle more efficient.

This chapter serves to introduce the topic of transistor device modeling for CAD. An overview of the concepts of bipolar and MOSFET transistor physics relevant to the subsequent discussion is followed by a brief description of various models for these transistors. Complete descriptions are not provided for the models, in the interest of brevity (references that exhaustively describe the models are provided for the interested reader); rather, the objective is to provide an overview of the salient features and shortcomings of the models, and the motivation for the advanced topics and techniques discussed in Chapters 3 and 4.

2.2 BIPOLAR JUNCTION AND HETEROJUNCTION BIPOLAR TRANSISTORS

Transistors may, in general, be divided into two classes: bipolar transistors and unipolar transistors. In bipolar transistors, both positive and negative free carriers take part in the operation of the device. In unipolar transistors, current is carried in the conducting channel only by the majority free carriers. Field effect transistors (FETs) are examples of unipolar transistors.

The bipolar junction transistor (BJT) is generally viewed as a current-controlled current source with a current gain β given by

$$\beta = \frac{\partial I_C}{\partial I_B} \tag{2.1}$$

where I_C is the collector current, and I_B is the base current. Alternatively, the device can be viewed as a voltage-controlled current source with a transconductance g_m given by

$$g_m = \frac{\partial I_C}{\partial V_{BE}} \tag{2.2}$$

The current gain β can be related to the physical parameters of the device by (assuming a short base and negligible base recombination)

$$\beta = \frac{D_B n_{iB}^2 W_E N_E}{W_B N_B D_E n_{iE}^2} \tag{2.3}$$

where D_B is the diffusion constant for electrons (in the case of an NPN device) in the base, D_E is that for holes in the emitter, W_B is the width of the emitter, W_E is that of the emitter, N_B is the doping concentration in the base, N_E is that in the emitter, n_{iB} is the instrinsic carrier concentration in the base, and n_{iE}, that in the emitter.

In order to achieve a high β a large N_E, a small value of N_B and a small W_B are desirable. However, the doping and the width of the base must be large enough to withstand the depletion occurring in the junctions, so that a punchthrough of the base does not occur. A higher base doping also allows a higher current density before high injection effects caused by the minority-carrier concentration reaching the doping level in the base occur. Further, a higher base doping and a larger base width give a lower base resistance, which increases the maximum frequency of oscillation f_{max}. Thus, it is quite difficult to achieve high gain, high f_{max}, and good breakdown characteristics simultaneously in a BJT device.

In the heterojunction bipolar transistor (HBT), a bandgap difference is introduced between the emitter and the base. The bandgap difference can be accomplished by using two different materials for the base and the emitter (for example, GaAs and AlGaAs, respectively) or by introducing a different material in the base, as in the SiGe HBT, where a strained layer of SiGe is grown. The former configuration is referred to as the *single heterojunction bipolar transistor* (SHBT), while the latter is known as the *double heterojunction bipolar transistor* (DHBT) since both the emitter and collector are of the same material. The result of either configuration is similar: β increases drastically with increasing bandgap difference. The reason for this is the strong dependence of the intrinsic carrier concentration on the bandgap difference:

$$n_i = \left(e^{-E_g/kT}\right)^{1/2} \tag{2.4}$$

where n_i is the intrinsic carrier concentration, E_g is the bandgap energy difference, k is Boltzmann's constant and T is the absolute temperature. Thus, β for the HBT becomes

$$\beta = \frac{D_B W_E N_E e^{\Delta E_g/kT}}{W_B N_B D_E} \tag{2.5}$$

where ΔE_g is the difference in bandgap between the emitter and the base. The exponential term in Equation 2.5 can be quite large, giving a large value of β. Furthermore, this gain enhancement can be achieved while simultaneously increasing the base doping. This means that the base width can be lowered without risking punchthrough or a high base resistance. The current density can also be increased, whereby the HBT can be made smaller than a comparable BJT, which reduces parasitics. As a result, both the unity current gain frequency f_T and the maximum frequency of oscillation f_{max}, which is the more important figure of merit for microwave applications, are much higher in an HBT than in a BJT.

2.3 BIPOLAR DEVICE MODELS

2.3.1 The Ebers–Moll Model

The first model to be proposed for the bipolar transistor was the Ebers–Moll (EM) model. It is a physical model, and in essence, it consists of two diodes in parallel with two current sources. The diodes represent the two junctions causing injection of carriers into the base. The current sources model how these carriers are swept through the depletion region of the other junction. The losses, for example, recombination in the base, are accounted for by the current transport factor α. Series resistors are added to the three terminals to represent the lead and contact resistances, and capacitors are added in parallel with the junctions to model the junction and parasitic capacitances. The earliest version of the EM model, the EM1 model [1], is only a DC model, and does not include charge storage. The next version, the EM2 model, introduced capacitances to model charge storage, in addition to adding base, emitter, and collector series resistances to make the DC model more accurate. The EM3 model expands on EM2 by adding base width modulation, β variation with voltage and current, and parameter variation with temperature. The Ebers–Moll static model, which forms the basis for most later bipolar transistor models, has two versions: the injection version and the transport version. They are mathematically identical, and differ only in the choice of reference currents. The transport version is preferred for computer simulation (the difference becomes more apparent at higher levels of model complexity) because with this topology, the reference currents remain ideal over many decades of current, both reference currents are completely specified by one constant, and description of the diffusion capacitances is easier [2]. The equivalent circuit diagram of a standard Ebers–Moll large-signal model is shown in Figure 2.1. The DC model is described by the diode equations

$$I_{CC} = I_S\left(e^{V_{BE}/V_T} - 1\right) \tag{2.6}$$

$$I_{EC} = I_S\left(e^{V_{BC}/V_T} - 1\right) \tag{2.7}$$

where I_S is the saturation current, V_T (= kT/q) is the thermal voltage, and β_F and β_R, the large-signal forward and reverse current gains of a common-emitter bipolar transistor, respectively, are given by

$$\beta_F = \frac{\alpha_F}{1-\alpha_F} \tag{2.8}$$

$$\beta_R = \frac{\alpha_R}{1-\alpha_R} \tag{2.9}$$

where α_F and α_R are the forward and reverse current gains in the common-base configuration, respectively. The two current sources I_{CC} and I_{EC} can be replaced by a single current source I_{CX} given by

$$I_{CX} = I_{CC} - I_{EC} \tag{2.10}$$

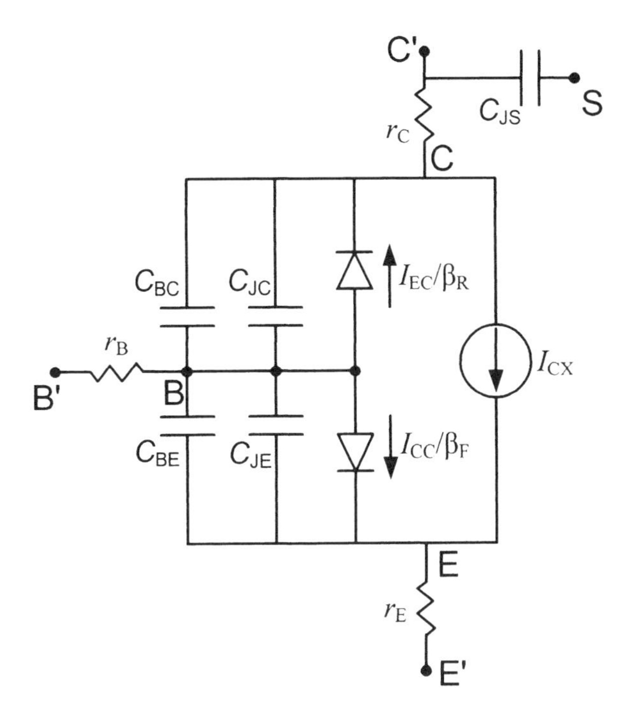

Figure 2.1. The Ebers–Moll large-signal model.

The terminal currents are, therefore,

$$I_B = \frac{I_{CC}}{\beta_F} + \frac{I_{EC}}{\beta_R} \tag{2.11}$$

$$I_C = I_{CX} - \frac{I_{EC}}{\beta_R} \tag{2.12}$$

$$I_E = \frac{I_{CC}}{\beta_F} - I_{CX} \tag{2.13}$$

2.3.2 The Gummel-Poon Model

The Gummel–Poon (GP) model [3], in its SPICE implementation called the *SPICE Gummel–Poon* (SGP) model, was by far the most widely used model for bipolar design until recently. The GP model treats the device from a more physics-based point of view compared to the engineeringwise very intuitive EM model. It includes effects that are not modeled in the basic EM model, important among which are the low current and the high injection effects. The GP model is built on the integral charge concept, which is the charge represented by the areal majority carrier density in the neutral emitter, base, or collector regions. Accordingly, the total majority charge in the neutral base region Q_B can written in the form

$$Q_B = Q_{B0} + C_{JE}V_{BE} + C_{JC}V_{BC} + \tau_{BF}I_{CC} + \tau_{BR}I_{EC} \tag{2.14}$$

where Q_{B0} is the zero-bias majority base charge, C_{JE} and C_{JC} are the base–emitter and base–collector junction capacitances, and τ_{BF} and τ_{BR} are the forward and reverse base transit times. The Gummel–Poon large-signal model is shown in Figure 2.2. As in the case of the EM model we define $I_{CX} = I_{CC} - I_{EC}$. The expressions for the diode currents I_{CC} and I_{EC} are given by

$$I_{CC} = \frac{I_{SS}}{q_b}\left(e^{V_{BE}/n_F V_T} - 1\right) \tag{2.15}$$

$$I_{EC} = \frac{I_{SS}}{q_b}\left(e^{V_{BC}/n_R V_T} - 1\right) \tag{2.16}$$

where I_{SS} replaces the I_S parameter of the EM model. This is because the pre-exponential parameter I_S is, in fact, not constant at high-level injection, so in the

Gummel–Poon model I_{SS} is defined at zero V_{BE} and V_{BC} (note that in the SGP model, I_{SS} and I_S coincide). n_F and n_R are the forward and reverse current emission coefficients, respectively. The normalized base charge q_b is defined as

$$q_B = \frac{Q_B}{Q_{B0}} \tag{2.17}$$

Using Equation 2.14, q_b can be expressed as

$$q_B = q_1 + \frac{q_2}{q_b} \tag{2.18}$$

where

$$q_1 = 1 + \frac{V_{BE}}{V_B} \tag{2.19}$$

$$q_2 = \frac{I_{SS}}{I_{KF}}\left(e^{V_{BE}/V_T} - 1\right) + \frac{I_{SS}}{I_{KR}}\left(e^{V_{BC}/V_T} - 1\right) \tag{2.20}$$

where V_A, the forward Early voltage, is defined as

$$V_A = \frac{Q_{B0}}{C_{JE}} \tag{2.21}$$

and V_B, the reverse Early voltage, is defined as

$$V_B = \frac{Q_{B0}}{C_{JC}} \tag{2.22}$$

The last two terms of Equation 2.14 are expressed in expanded form to give Equation 2.19. The parameters I_{KF} and I_{KR} in Equation 2.20 are the "knee" currents corresponding to the $\ln(I_C)$–V_{BE}, and the $\ln(I_B)$–V_{BC} curves, respectively, and are related to the parameters of Equation 2.14 as

$$I_{KF} = \frac{Q_{B0}}{\tau_{BF}} \tag{2.23}$$

$$I_{KR} = \frac{Q_{B0}}{\tau_{BR}} \tag{2.24}$$

The parameter q_1 models base-width modulation, as is apparent on observing that it is unity at zero bias, reflecting the fact that the physical and electrical base widths are equal at zero bias; greater than unity when both junctions are forward-biased, because the electrical base width is greater than the physical base width; and less than unity if both junctions are reverse-biased, since then the physical base width exceeds the electrical base width. The parameter q_2 models high-level injection effects. Under low-level injection, that is, if $q_2 << q_1^2/4$, solving the quadratic Equation 2.19 for q_b, we get

$$q_b \approx q_1 \tag{2.25}$$

Similarly, for the case of high-level injection ($q_2 >> q_1^2/4$), we get

$$q_b \approx \sqrt{q_2} \tag{2.26}$$

Using Equation 2.25 to solve for I_C in the normal active region at high-injection levels, with the additional assumption that $V_{BC} = 0$ for simplicity, we get

$$I_C \approx \sqrt{I_{SS} I_{KF}}\, e^{V_{BE}/2V_T} \tag{2.27}$$

which models the familiar reduction in slope of the $\ln(I_C)$–V_{BE} curve by half, at high injection levels.

The Gummel–Poon model models the variation of current gain β with current. The reduction in the current gain at high current is seen from the discussion on high-level injection above. The Gummel–Poon model also models the drop in current gain at low current, by including additional components in the base current I_B to model the recombination of carriers at the surface, the recombination of carriers in the base–emitter space charge layer, and the formation of emitter–base surface channels [2]. Since all three components exhibit a similar relationship to the base–emitter voltage V_{BE}, they can be modeled by a composite current, I_{BEn}, given by [2]

$$I_{BEn} = C_2 I_{SS}\left(e^{V_{BE}/n_{EL}V_T} - 1\right) \tag{2.28}$$

where C_2 is the forward low-current nonideal base current coefficient and n_{EL} is the low-current forward region base–emitter emission coefficient. Similarly a composite current I_{BCn} is defined to describe the additional current component when the base–collector junction is forward biased:

$$I_{BCn} = C_4 I_{SS}\left(e^{V_{BC}/n_{CL}V_T} - 1\right) \tag{2.29}$$

where C_4 and n_{CL} are the reverse low-current nonideal base current coefficient and the low-current base–collector emission coefficient, respectively. Thus, the general expression for base current becomes

$$I_{B} = \frac{I_{CC}}{\beta_{F}} + \frac{I_{EC}}{\beta_{R}} + I_{BEn} + I_{BCn} \tag{2.30}$$

Note that β_F and β_R in this equation refer to the current gain in the midcurrent region, where current gain is treated as being constant.

The Gummel–Poon base resistance model takes into account the effect of current crowding, as a result of which base resistance becomes dependent on current. The base resistance model is expressed as

$$r_{b} = r_{bm} + 3\left(r_{b0} - r_{bm}\right)\left(\frac{\tan z - z}{z \tan^{2} z}\right) \tag{2.31}$$

where r_{bm} is the minimum base resistance that occurs at high currents, r_{b0} is the base resistance at zero bias, and z is given by

$$r_{b} = \frac{-1 + \left[1 + (12/\pi)^{2}\left(I_{B}/I_{RB}\right)\right]^{1/2}}{\left(24/\pi^{2}\right)\left(I_{B}/I_{RB}\right)^{1/2}} \tag{2.32}$$

where I_{RB} is the current at which the base resistance is half its maximum value.

The Gummel–Poon model has a distributed base–collector capacitance, as seen from Figure 2.2. A parameter X_{JC}, which takes values between 0 and 1, is used to partition the base–collector junction capacitance C_{JC} between the internal base–collector junction capacitance ($C_{JCi} = X_{JC}C_{JC}$) and the external base–collector capacitance ($C_{JCx} = C_{JC} - X_{JC}C_{JC}$). Note that the junction or space-charge capacitances are modeled as

$$C_{Jk} = \frac{C_{Jk0}}{\left(1 - \frac{V_{Bk}}{V_{Jk}}\right)^{m_{Jk}}}, \quad \text{for } V_{Bk} < FC.V_{Jk}, \; k = C, E \tag{2.33}$$

and continue as a linear extrapolation for $V_{Bk} > FC.V_{Jk}$. Here C_{Jk0} are the capacitances at zero bias, V_{Jk} are the (base-–collector and base–emitter) barrier potentials, m_{Jk} are gradient coefficients that usually take values between 0.3 and 0.5, and FC is a factor between 0 and 1 used to indicate the voltage in the forward-bias region beyond which the capacitance is modeled by a linear extrapolation.

The diffusion capacitances C_{BC} and C_{BE} are obtained from the corresponding stored charges Q_{BE} and Q_{BC}, which are modeled by the equations:

$$Q_{BE} = \tau_{FF} I_{CC} \tag{2.34}$$

$$\tau_{FF} = \tau_F \left[1 + X_{\tau F} \left(\frac{I_{CC}}{I_{CC} + I_{\tau F}} \right)^2 \cdot e^{V_{BC}/1.44 V_{\tau F}} \right] \tag{2.35}$$

$$Q_{BC} = \tau_R I_{EC} \tag{2.36}$$

where τ_F and τ_R are the ideal forward and reverse transit times, respectively; τ_{FF} is the modulated forward transit time; $X_{\tau F}$ is a parameter that controls the total falloff of cutoff frequency f_T; $V_{\tau F}$ is a parameter that determines the change in f_T with V_{CE}; and $I_{\tau F}$ controls the change in f_T with respect to current. These equations model the variation in τ_F (and hence f_T) with I_C, and empirically capture the decrease in f_T at high currents due to effects such as base pushout and quasisaturation. From the preceding formulation for stored charge, the diffusion capacitances are easily computed as

$$C_{BE} = \frac{dQ_{BE}}{dV_{BE}} \tag{2.37}$$

$$C_{BC} = \frac{dQ_{BC}}{dV_{BC}} \tag{2.38}$$

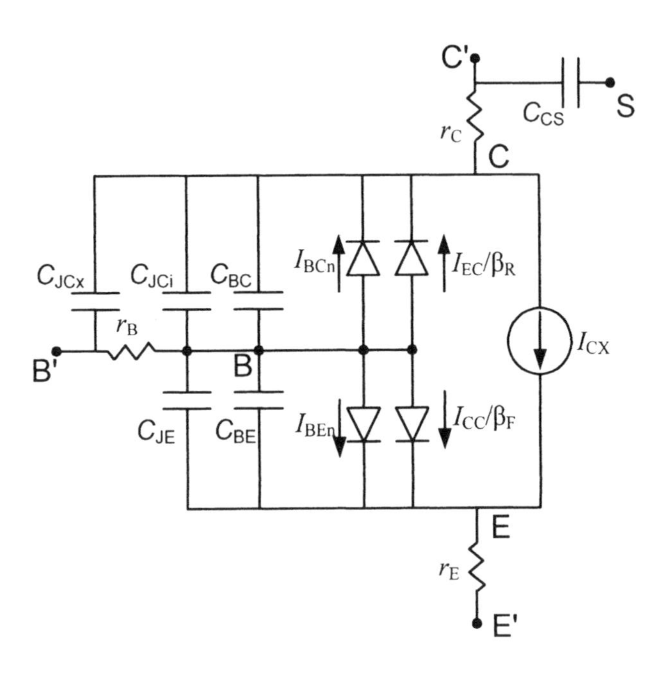

Figure 2.2. The Gummel–Poon large-signal model.

Excess phase, which refers to the excess phase shift of the measured forward transverse current compared to that predicted by the pole, because the lumped-element model with finite poles and zeros only approximates the distributed phenomena in the base, is corrected by inserting extra phase delay in the model.

The Gummel–Poon model, in its SPICE form (the SGP model), remained the standard model for bipolar circuit simulation until recently. However, it suffers from limitations, which became increasingly apparent with the advancement of bipolar device technology and the consequent application of bipolar devices to high-frequency circuit design. It is worth noting here that the standard SGP model is actually a simplified implementation of the original Gummel–Poon model, and varies from it in some respects, especially in the modeling of the Early effect. This is because, in the SGP model, the parameter q_1 is not modeled according to the original Gummel–Poon model formulation of Equation 2.19. Instead the approximation of Equation 2.39 is used:

$$q_1 \approx \frac{1}{1 - V_{BE}/V_B - V_{BC}/V_A} \tag{2.39}$$

Further, q_b is approximated as

$$q_b = \frac{q_1}{2}\left(1 + \sqrt{1 + 4q_2}\right) \tag{2.40}$$

The simplified SGP Early effect model predicts the output conductance inaccurately, and collector resistance is treated as constant. Important physical phenomena like self-heating and avalanche multiplication are ignored. The modeling of high-frequency and temperature effects is inadequate. There is no substrate model. These limitations pose serious problems in predicting the performance of modern bipolar devices with sufficient accuracy. As a result, improved models have been developed for bipolar devices, the first of which to be discussed here is the VBIC model.

2.3.3 The VBIC Model

The VBIC (vertical bipolar inter-company) model, first released under the name VBIC95 [4], includes several improvements over the SGP model. As indicated by its name, it focuses on a vertical transistor structure, which is the structure of most high-frequency bipolar devices today, since the base can be made very thin using a vertical structure, resulting in high f_T. The VBIC model includes an intrinsic transistor based on the Gummel–Poon formulation, and an extrinsic parasitic transistor modeled using a partial Gummel–Poon model. The intrinsic and extrinsic transistors are typically NPN and PNP BJTs, respectively. The equivalent circuit of the VBIC model is shown in Figure 2.3. An important difference in the current source model of VBIC compared to SGP is that the base current is not linked to the collector current by a β parameter, reflecting the fact that the physical mechanisms that control the two are separate [4]. The VBIC current source models are given by

$$I_{CC} = \frac{I_F - I_R}{q_b} \tag{2.41}$$

$$I_F = I_S\left(e^{V_{BE}/n_F V_T} - 1\right) \tag{2.42}$$

$$I_R = I_S\left(e^{V_{BC}/n_R V_T} - 1\right) \tag{2.43}$$

$$I_{BE} = I_{BEI}\left(e^{V_{BE}/n_{EI} V_T} - 1\right) + I_{BEN}\left(e^{V_{BE}/n_{EN} V_T} - 1\right) \tag{2.44}$$

where the I_{BEI} and I_{BEN} terms of Equation 2.44 represent the ideal and nonideal components of the b–e component of the base current. Typically, n_{EI} and n_{EN} are of the order of 1 and 2, respectively. $I_{BE,tot}$ is partitioned between I_{BE} and I_{BEx} as W_{BE} and $(1-W_{BE})$, respectively, to model the distributed nature of the base, where W_{BE} is a parameter that takes values between 0 and 1. The b–c component of the base current I_{BC} includes ideal and nonideal components similar to the b–e component:

$$I_{BC} = I_{BCI}\left(e^{V_{BC}/n_{CI} V_T} - 1\right) + I_{BCN}\left(e^{V_{BC}/n_{CN} V_T} - 1\right) \tag{2.45}$$

Avalanche multiplication is modeled by the addition of a weak avalanche current to the b–c component of the base current [5]

$$I_{gc} = \left(I_{CC} - I_{BC}\right) \cdot A_{VC1} \cdot \left(P_C - V_{BC}\right) \cdot \exp\left(- A_{VC2}\left(P_C - V_{BC}\right)^{M_C - 1}\right) \tag{2.46}$$

where P_C and M_C are the built-in potential and grading coefficient of the b–c junction, respectively, and A_{VC1} and A_{VC2} are model parameters. The modeling of the normalized base charge q_b is improved compared to SGP, in that q_1 and q_b revert to Equation 2.19, and the nonnegative root of Equation 2.18, given by Equation 2.47, respectively. This results in an improved Early effect model in VBIC compared to SGP:

$$q_b = \frac{q_1}{2} + \sqrt{\frac{q_1^2}{4} + q_2} \tag{2.47}$$

The depletion capacitance model formulation is similar to the piecewise model of SGP. In VBIC, the base-emitter junction capacitance C_{JE} is partitioned into C_{JEi} and C_{JEx}, as W_{BE} and (1- W_{BE}), respectively. An alternate single-piece junction capacitance model, in which the capacitance smoothly limits to a constant value for junction biases greater than the built-in potential, and that can be chosen instead of the piecewise model, is also available in VBIC [4]. The diffusion capacitance model is enhanced with respect to SGP, with the addition of the parameter q_1 in the expression for τ_{FF}, the modulated forward transit time, as shown below:

$$Q_{BE} = \tau_{FF} I_F \tag{2.48}$$

$$\tau_{FF} = \tau_F \left(1 + Q_{\tau F}\, q_1\right)\left[1 + X_{\tau F}\left(\frac{I_F}{I_F + I_{\tau F}}\right)^2 \cdot e^{V_{BC}/1.44 V_{\tau F}}\right] \tag{2.49}$$

$$Q_{BC} = \tau_R I_R \tag{2.50}$$

Another feature of the VBIC model is the modeling of the collector resistance, which consists of a constant extrinsic resistance R_{Cx} and a variable intrinsic resistance R_{Ci} to model quasisaturation. Quasisaturation refers to an effect that occurs at high currents, where, because of a large internal voltage drop in the collector region, the base–collector junction might become forward-biased although the external base–collector voltage remains in reverse bias. This internal voltage drop occurs in the active part of the transistor, under the emitter. The forward biasing of the base–collector junction results in an increase of the reverse current and a charge storage in the collector epilayer (Q_{epi}), which must be added to the already existing base charge Q_B. This increase in Q_B, often called *base widening* or *base pushout*, reduces the collector current and current gain. Recombination of the stored charge Q_{epi} increases the base current and further reduces current gain. Q_{epi} also causes an increase in transit time, hence reducing the cutoff frequency. The expression for the current in the intrinsic collector is a modification of the Kull model [6]:

$$I_{Rci} = \frac{I_{epi0}}{\sqrt{1 + \left(\dfrac{I_{epi0} R_{Ci}}{V_0\left(1 + \dfrac{\sqrt{0.01 + V_{Rci}^2}}{2 V_0 H_{RCF}}\right)}\right)^2}} \tag{2.51}$$

$$I_{epi0} = \frac{1}{R_{Ci}}\left[V_{Rci} + V_T\left(K_{bci} - K_{bcx} - \ln\frac{1 + K_{bci}}{1 + K_{bcx}}\right)\right] \tag{2.52}$$

$$K_{\mathrm{bci}} = \sqrt{1 + \gamma \exp\left(\frac{V_{\mathrm{BCi}}}{V_{\mathrm{T}}}\right)} \tag{2.53}$$

$$K_{\mathrm{bci}} = \sqrt{1 + \gamma \exp\left(\frac{V_{\mathrm{BC}}}{V_{\mathrm{T}}}\right)} \tag{2.54}$$

where $V_{\mathrm{Rci}} = V_{\mathrm{BCi}} - V_{\mathrm{BCx}}$, $V_0 = w_{\mathrm{epi}} v_{\mathrm{sat}} / \mu_0$, and H_{RCF} is a parameter to account for the increase in collector current with increasing V_{Rci} at high V_{BCi}. Additionally, Q_{BC} includes a component $Q_{\mathrm{BCq}} = Q_{\mathrm{C0}} K_{\mathrm{bci}}$, which models part of the charge associated with base pushout, and the other part is $Q_{\mathrm{BCx}} = Q_{\mathrm{C0}} K_{\mathrm{bcx}}$ [4].

The base resistance in VBIC is modulated by q_{b}, a modification from the empirical formulation of SGP.

$$R_{\mathrm{B}} = R_{\mathrm{Bx}} + \frac{R_{\mathrm{Bi}}}{q_{\mathrm{b}}} \tag{2.55}$$

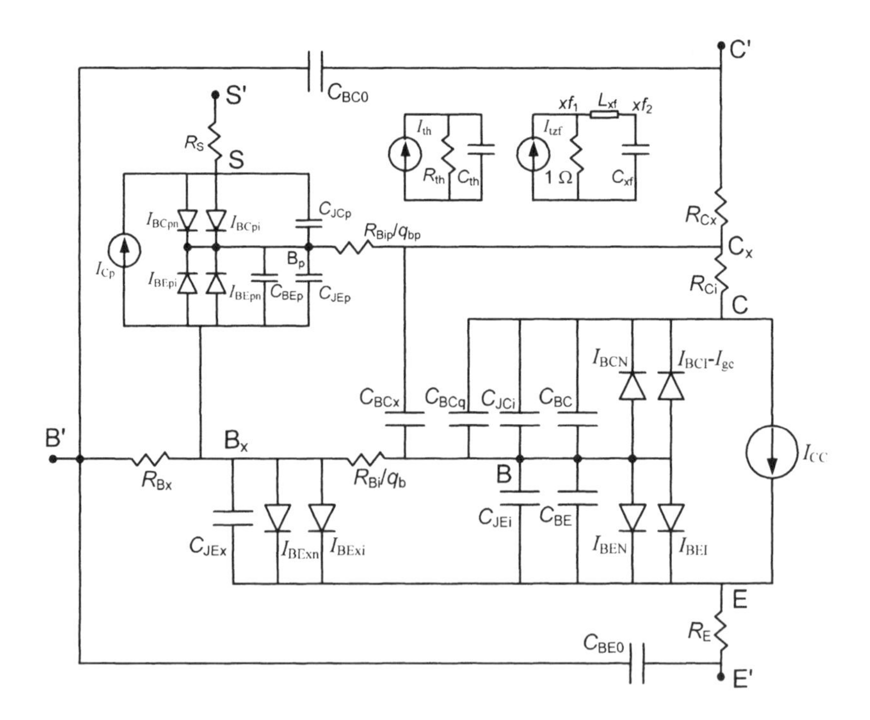

Figure 2.3. Equivalent circuit for the VBIC model.

The emitter and substrate resistances, R_E and R_S, respectively, are constant. Parasitic capacitors C_{BE0} and C_{BC0} are included to model constant *b–e* and *b–c* overlap capacitances. The excess phase effect is modeled by a subcircuit through the parameter T_D which reflects the excess phase shift. The formulation is consistent between AC and transient analysis. A subcircuit is also added to model self-heating, as seen from Figure 2.3. The substrate model is complex in contrast with its absence from SGP, and consists of a parasitic substrate transistor, which is modeled by a simplified Gummel–Poon model.

In summary, the VBIC model offers several improvements over the SGP model; the modeling of base width modulation is more accurate in VBIC, a modified Kull model is included to model quasisaturation, a parasitic substrate transistor is introduced, a weak avalanche current model is included, static temperature mapping is extensive, a subcircuit to model self-heating model is introduced, a first-order approximation for the distributed base is included, the space charge capacitance model is improved, overlap capacitances are included, the transit time model is enhanced, and the treatment of excess phase is consistent between AC and transient analysis.

2.3.4 MEXTRAM

MEXTRAM (most exquisite transistor model) [7,8] is a bipolar model developed by Philips, and made available in the public domain. MEXTRAM has some significant differences from the Gummel–Poon and VBIC model formulations. For instance, in the MEXTRAM formulation of the forward and reverse collector currents (I_F and I_R), the nonideality factors or emission coefficients (n_F and n_R) are not present. This is because, in most Si-based processes, these values are close to unity. Any nonideality in the collector current is due mainly to the reverse Early effect, and is accounted for in the model for the normalized base charge. The basic expressions relating to the main collector current model are given by

$$I_{CC} = \frac{I_F - I_R}{q_b} \tag{2.56}$$

$$q_b = q_1(1 + q_2) \tag{2.57}$$

$$q_1 = 1 + \frac{Q_{tE}}{Q_{B0}} + \frac{Q_{tC}}{Q_{B0}} \tag{2.58}$$

$$q_2 = \frac{Q_{BE}}{Q_{B0}} + \frac{Q_{BC}}{Q_{B0}} \tag{2.59}$$

where $Q_{tE} = C_{JE0}V_{tE}$ and $Q_{tC} = C_{JC0}V_{tC}$, where V_{tE} and V_{tC} describe the curvature (but not the magnitude) of the depletion charges as a function of the junction biases. The Early effect modeling is different in MEXTRAM compared to Gummel–Poon, since q_1 is modeled differently. In MEXTRAM, high injection, described by q_2, is related to the diffusion charges Q_{BE} and Q_{BC}. Q_{B0} is the base charge at zero bias. The normalized diffusion charges can be expressed in terms of the normalized electron densities at the edges of the neutral base region, n_0 and n_B, to give

$$q_2 = n_0/2 + n_B/2 \tag{2.60}$$

where

$$n_0 = \frac{4I_F/I_K}{1+\sqrt{1+4I_F/I_K}} \tag{2.61}$$

$$n_B = \frac{4I_R/I_K}{1+\sqrt{1+4I_R/I_K}} \tag{2.62}$$

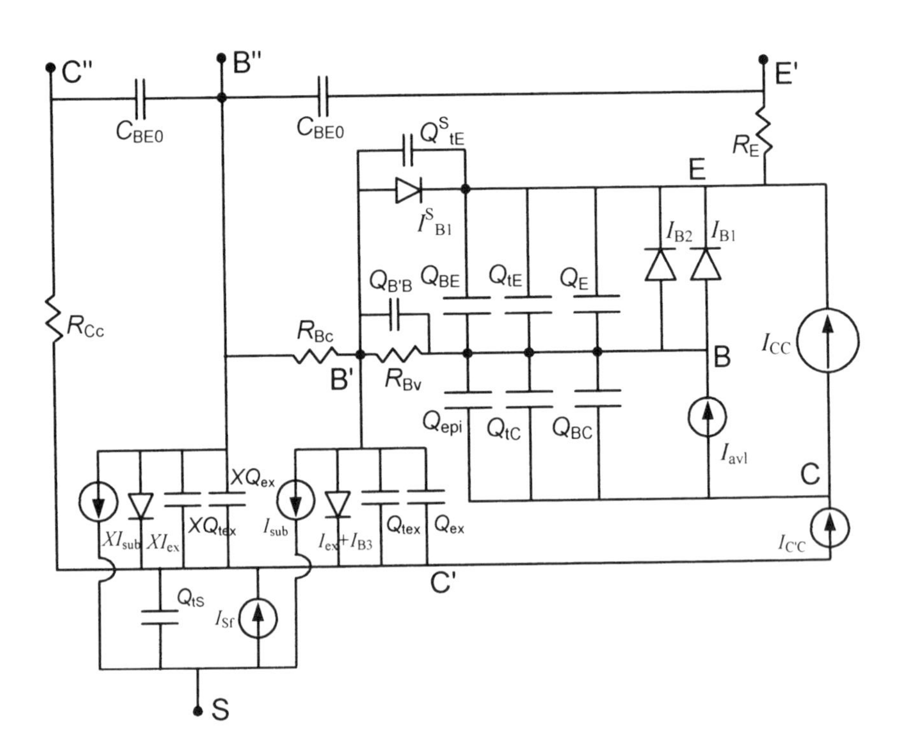

Figure 2.4. Equivalent circuit for MEXTRAM.

MEXTRAM uses only one knee current I_K unlike Gummel–Poon. The forward base current consists of an ideal component and a nonideal recombination-based component. The reverse base current consists of three components: electrons injected from the collector into the extrinsic base, a nonideal Shockley-Read-Hall (SRH) recombination-based component, and the substrate current of the parasitic transistor. MEXTRAM also includes an avalanche current model, and the avalanche current is a function of bias and current. MEXTRAM is capable of modeling snapback effects at high currents, as an option. It also contains a substrate current, which describes the main current of the parasitic PNP. In MEXTRAM, the parasitic PNP transistor may be viewed as consisting of a sidewall and a floor component. Figure 2.4 shows the equivalent circuit for MEXTRAM.

The model for the depletion capacitances resembles the Gummel–Poon formulation. However MEXTRAM does not linearize the capacitances above a certain forward bias, rather it uses a smooth transition to a constant capacitance. Since the transition happens in a region where the diffusion capacitances dominate, this difference is not very important. MEXTRAM partitions the base–emitter depletion capacitance into two capacitances, as does VBIC. MEXTRAM has quite a sophisticated model for the base–collector depletion capacitance, and includes current modulation effects. It takes into account the finite thickness of the epilayer and the influence of the collector current. MEXTRAM has constant overlap capacitances, similar to VBIC.

The modeling of the diffusion charges in MEXTRAM has contributions from the base charge (Q_{BE}) and charge stored in the emitter (Q_E) to the low-current transit time:

$$Q_{BE} = q_1 \tau_B I_F \frac{2}{1 + \sqrt{1 + 4I_F / I_K}} \tag{2.63}$$

$$Q_E = \tau_E I_S e^{V_{BE}/m_\tau V_T} \tag{2.64}$$

where the low-current transit time is the sum of τ_B and τ_E, and m_τ is a nonideality factor. At higher currents, since MEXTRAM has a collector epilayer model, the intrinsic base–collector junction can become forward-biased and the reverse part of the main current I_R is no longer negligible compared to the forward part I_F. The resulting extra charge is modeled as follows:

$$Q_{BC} = q_1 \tau_B I_R \frac{2}{1 + \sqrt{1 + 4I_R / I_K}} \tag{2.65}$$

The charge in the collector epilayer due to base widening is given by

$$Q_{\mathrm{epi}} \approx \tau_{\mathrm{epi}} I_{\mathrm{epi}} \left(\frac{x_{\mathrm{i}}}{W_{\mathrm{epi}}} \right)^2 \tag{2.66}$$

where x_i is the extent of base widening, W_{epi} is the epilayer thickness, and I_{epi} is the epilayer current.

The base resistance in MEXTRAM, as in VBIC, is partitioned into a constant and variable part. The model for the variable base resistance in MEXTRAM considers both conductivity modulation by q_b and DC emitter current crowding. In addition, AC current crowding is modeled by a capacitance in parallel with the variable base resistance. The model for the epilayer resistance in MEXTRAM, as in VBIC, is based on the Kull formulation [6]. When velocity saturation is excluded, the value of the epilayer resistance predicted by MEXTRAM is the same as that in VBIC. The effect of velocity saturation in the collector epilayer is that it extends the quasisaturation regime to lower collector currents and higher collector voltages. The onset of quasisaturation is reached at a lower collector current with the MEXTRAM model, compared to the VBIC model.

MEXTRAM models intrinsic excess phase shift by partitioning the base charge (when the model parameter *EXPHI* =1); that is, Q_{BC} and Q_{BE} are changed from their previously computed values to ($Q_{BE}/3$ + Q_{BC}) and $2Q_{BE}/3$, respectively. The effect of AC current crowding is also included, by the addition of extra charge in parallel to the intrinsic base resistance.

More details regarding the MEXTRAM model can be found in references [7–9].

2.3.5 HICUM

HICUM (high current model) is a bipolar transistor model developed for vertical BJTs and HBTs. It was initially developed with emphasis on modeling the high-current region of operation [10,11], which is important in certain high-speed applications. The model was later expanded to model high-frequency analog operation [12]. HICUM is based on the generalized integral charge-control relation (GICCR) [13], from which the transfer current I_T is modeled as

$$I_{\mathrm{T}} = I_{\mathrm{TF}} - I_{\mathrm{TR}} = \frac{C_{10}}{Q_{\mathrm{PT}}} \left(e^{V_{\mathrm{BE}}/V_{\mathrm{T}}} - e^{V_{\mathrm{BC}}/V_{\mathrm{T}}} \right) \tag{2.67}$$

where C_{10} is the GICCR constant and Q_{PT} is a weighted hole charge expressed as [14]

$$Q_{\mathrm{PT}} = Q_{\mathrm{P0}} + h_{\mathrm{jEi}} Q_{\mathrm{jEi}} + h_{\mathrm{jCi}} Q_{\mathrm{jCi}} + Q_{\mathrm{F}} + Q_{\mathrm{R}} \tag{2.68}$$

where Q_{P0} is the zero-hole charge, Q_{jEi} and Q_{jCi} are the base–emitter and base–collector depletion charges of the internal transistor, respectively, h_{jEi} and h_{jCi} are the weighting factors for HBTs, and Q_F and Q_R are the diffusion charges, which are calculated from the integral of the forward and reverse transit time, respectively. The expression for the bias dependence of the forward transit time is

$$\tau_F = \tau_{F0} + \Delta\tau_F \tag{2.69}$$

where τ_{F0} is the low-current forward transit time and $\Delta\tau_F$ models the increase in transit time at medium and high current densitites. Models such as Gummel–Poon also apply the ICCR concept, but in HICUM, (G)ICCR is applied without simplifications and additional fitting parameters. Treating parameters that determine dynamic behavior of the transistor, such as depletion capacitances and transit time of carriers and the associated charges, as basic quantities of the model, and accurately approximating them as a function of bias, results in an accurate description of small-signal, dynamic large-signal, and via the GICCR, also DC behavior [15]. This coupling between static and dynamic descriptions results in a reduction of fitting parameters like Early voltages and knee currents, and simplifies model parameter extraction. In HICUM, all base currents are modeled independently of the transfer current to reflect the physically independent mechanisms governing base and collector currents. HICUM features a weak avalanche current model to indicate the onset of breakdown.

HICUM contains an analytical formulation for the bias dependence of the variable part of the base resistance that includes emitter crowding and is valid up to very high current densities. A shunt capacitance across the variable base resistance also accounts for the frequency dependence of current crowding in small-signal operation, as in MEXTRAM. Also, similar to MEXTRAM, for the depletion capacitance model, HICUM uses a smooth transition to a constant value above a certain forward bias. HICUM also accounts for the possible reachthrough of the base–collector and collector–substrate depletion layer by switching the capacitance model to a constant value if the reverse bias exceeds a model parameter defined critical value. Nonquasistatic (NQS) effects occurring at high frequencies, which result in the delayed reaction of the minority charge Q_F, and transfer current I_T, with respect to the junction voltage, are accounted for by introducing additional delays for Q_F and I_T, which are modeled as a function of bias by relating them to the transit time τ_F by means of model parameters. Additional time delay, which results in excess phase, is modeled using a second-order Bessel polynomial for both AC and transient analysis in order to maintain consistency between the respective results. Self-heating effects are modeled in HICUM by means of a sub-circuit, as in VBIC. In addition to modeling high-current effects, including quasisaturation, emitter current crowding, emitter periphery injection, two- and three-dimensional collector current spreading, NQS effects, and avalanche multiplication, HICUM also models bandgap differences found in HBTs, and tunneling in the base–emitter junction. It features a distributed high-frequency model for the external base–collector and includes a parasitic PNP model. The elements of the HICUM model have been made a function of transistor configuration, which results in a sophisticated set of scaling equations due to the large variety of vertical device designs and lateral configurations used in circuit design. More detailed descriptions of HICUM may be found in references [15,16].

The advanced version of HICUM, namely HICUM/L2 (Level 2), is a very detailed description of the physics of the device and is quite complex. The equivalent circuit of HICUM/L2 is shown in Figure 2.5. In many practical circuit applications, using Level 2 is overkill, and computational effort is unnecessarily expended in using such a complex model for circuit simulation. A simpler model is often adequate to predict the circuit characteristics with sufficient accuracy. Hence, a simplified version called HICUM/L0 (Level 0) [17–19] has been developed, which is more accurate than SGP, but much less complex than HICUM/L2.

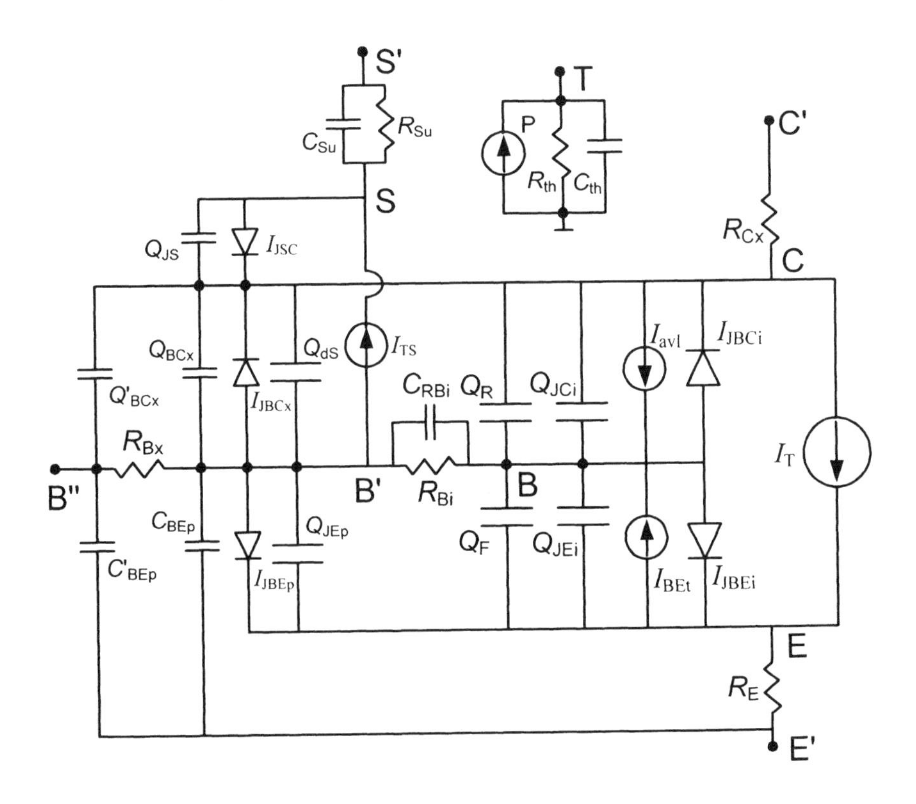

Figure 2.5. Equivalent circuit for HICUM/Level2.

2.4 MOSFET DEVICE PHYSICS

In this section, aspects of MOSFET device physics relevant to device modeling are reviewed. The simplest approach to describing the conducting channel of a MOSFET is to treat the inversion layer as uniform for $x < x_i$ and nonexistent for $x > x_i$, as shown in Figure 2.6. This leads us to the basic equations relating the electrical parameters associated with this structure, assuming that the conducting layer is an infinitesimally thin 2D charge layer:

$$dV_y = \frac{I_{ds}}{\mu W Q_{inv}(y)} dy \tag{2.70}$$

$$Q_{inv} = -C_{ox}\left[V_{gs} - V_{fb} - V(y) - 2\varphi_f\right] - \gamma\left[V(y) + 2\varphi_f + V_{sb}\right]^{1/2} \tag{2.71}$$

where I_{ds} is the channel current, dV_y is the potential difference across a small sliver of the channel dy at position y along the channel, μ is the mobility, W is the FET width, $Q_{inv}(y)$ is the inversion charge as a function of y, C_{ox} is the oxide capacitance, V_{gs} is the gate-source voltage, V_{fb} is the flatband voltage, φ_f is the Fermi potential, V_{bs} is the bulk–source voltage, and γ is given by

$$\gamma = \frac{\sqrt{2\varepsilon_s q N_A}}{C_{ox}} \tag{2.72}$$

where q is the electron charge, ε_s is the permittivity of silicon, and N_A is the substrate doping concentration. As the drain bias of a MOSFET is increased, keeping the gate bias fixed (with $V_{gs} > V_{th}$), for low drain bias, the drain current first increases with the drain bias. This region of operation is called the *linear region.* Beyond a certain drain bias, known as the *saturation voltage* (V_{dsat}), the drain current remains constant, and no longer increases with increasing drain bias. This is called the *saturation region.* For low drain biases, there is an inversion layer connecting the source and drain, which behaves as a resistor. For large drain bias voltages greater than the saturation voltage, the drain field prevents the inversion layer from reaching the drain diffusion, the channel pinches off, and the drain current becomes independent of drain bias. According to this simple FET analysis, below the threshold voltage, the inversion charge is reduced to zero, and the drain current is zero. In reality, there is a current in this region of operation, called the subthreshold region, which can be expressed most generally as

$$I_{ds} \propto e^{V_{gs}} \tag{2.73}$$

A more rigorous analysis of the FET drain current shows that when $V_{gs} > V_{th}$, current transport is dominated by drift due to the lateral electric field, while below the threshold voltage, it is dominated by diffusion of carriers over the channel potential barrier. Discontinuities in the analytical models for the current in different regions of operation are handled by introducing empirical curve-fitting expressions.

The MOSFET contains many capacitances, which have to be modeled. These are capacitances between source and substrate, drain and substrate, and between the gate and the rest of the device. The latter are usually the largest capacitive load introduced by the FET in a circuit, and are bias-dependent. In the zero-bias condition, there are small capacitances between the gate and the other terminals. In the linear region, the inversion layer is electrically connected to the source and drain, and screens the bulk from the gate, removing or reducing this component of the capacitance. Also, the capacitance between the gate and source or drain is now that between the gate and the inversion layer, which becomes the dominant intrinsic capacitance. In the saturation region, the inversion layer is not electrically connected to the drain, so the gate–drain capacitance becomes small, but the gate–source capacitance remains large. To model the gate capacitance, the charge on the gate (and on other terminals) must be found as a function of the bias voltages, keeping in mind the requirement of conservation of charge.

The early FET models were developed for what have now come to be known as "long-channel FETs." However, with the evolution of CMOS technology, driven by the need to achieve greater speed of operation with higher levels of integration, FET dimensions have decreased and the device structure has become more complicated. A new class of effects, called "small-geometry effects," emerged as being important to sufficiently represent the electrical behavior of the device. Often, these effects are added as corrections, with empirical formulations, to the original long-channel FET models.

In a long-channel FET, the depletion regions corresponding to the drain–bulk and source–bulk P–N junctions and the depletion region of the MOS capacitor, may simply be represented by the two-terminal equations for these depletion regions. In a short-channel FET, the overlap of the P–N junction depletion regions with the gate depletion region becomes significant. So the total depletion charge, rather than being merely the sum of the depletion charges from these three depletion regions, is reduced from that value. The result is that the inversion charge will be larger than that predicted by simple long-channel models. As the channel length decreases further, the source and drain depletion regions can themselves overlap. Thus, charge sharing occurs between the drain and the source, which further decreases the actual depletion charge (and increases the inversion charge) from the prediction of a long-channel model. Further, the extent of charge sharing varies with drain bias, since the extent of the drain depletion region changes with drain bias.

The mobility μ which is classically defined as the ratio of the carrier velocity v and the lateral electric field E, is treated as constant in long-channel FET models. However, as gate oxide thickness decreases, and the vertical field at the silicon surface increases, using a constant value for mobility is no longer valid. In fact, mobility decreases with increasing gate field. In addition to this, the ohmic definition of mobility as the ratio of carrier velocity and lateral electric field itself becomes invalid for large values of lateral field. The carrier velocity deviates from this linear relationship and becomes constant at high field values, a phenomenon known as *carrier velocity saturation*. The constant velocity at high field is called *saturation velocity*. Further, in reality, as the lateral field varies along the channel, the mobility also varies along the channel. The effect of carrier velocity saturation is that, in short-channel FETs, the drain current saturates below the expected V_{dsat} value. This effect is usually modeled by using carrier velocity saturation to describe the reduction in mobility, which is then used to arrive at a reduced value for the saturation voltage, which predicts the lower saturated drain current value.

An important characteristic of short-channel MOSFETs is channel length modulation. This reflects the fact that drain current is not constant above the saturation voltage in short-channel FETs. This is because, for $V_{ds} > V_{dsat}$, the pinchoff point moves away from the drain toward the source, a phenomenon that is negligible in a long-channel FET but not in a short-channel device. As the pinchoff point moves towards the source, the lateral field along the channel increases, which increases the drain current. Thus the drain current becomes dependent on the drain bias voltage, in saturation, in a short-channel device.

Drain-induced barrier lowering (DIBL) is another effect that manifests itself in short-channel devices. In a MOSFET, there is a barrier between the diffusion regions and the channel, due to the built-in junction potential. In a long-channel FET, when a drain bias is applied, only the immediate vicinity of the drain is affected, and the effect on the channel potential profile is negligible. This is not the case in a short-channel FET, where a drain bias changes the potential profile along the channel, and lowers the barrier at the source–substrate junction. This effect is called drain induced barrier lowering, its impact being that it allows the carriers to traverse the channel at a gate voltage lower than expected. Thus, DIBL decreases the threshold voltage of the FET.

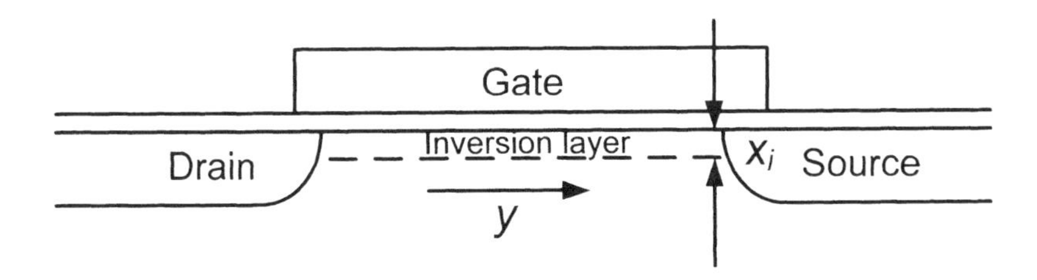

Figure 2.6. Simple description of MOSFET cross section.

As MOSFET channel length has decreased with CMOS technology scaling, the gate oxide thickness has also been reduced, to minimize short-channel effects. A consequence of the scaling of oxide thickness to very small dimensions is that the electrical barriers in the device begin to lose their insulating properties, and the oxide becomes subject to direct band-to-band tunneling, which results in a gate leakage current, which increases the standby power of ICs.

Gate-induced drain leakage is a phenomenon observed in advanced CMOS technologies. Considering an N-channel MOSFET in such a technology, when V_{gd}, the gate–drain bias, is negative, an electric field can develop across the thin oxide that is sufficiently high to deplete the N-type drain in the overlap region near the silicon surface. Electrons can tunnel from the valence band into the conduction band, generating holes. This minority-carrier generation can also occur by trap-assisted tunneling. The generated holes are swept into the substrate by the lateral field, and the electrons into the drain, resulting in gate-induced drain leakage.

Another effect caused by increasing field strengths in the channel, as CMOS technology has scaled to smaller dimensions, is hot-carrier degradation. The velocity of the carriers in a FET channel show a distribution, and this distribution becomes wider as the field strength increases with decreasing channel lengths. High-energy carriers at the high end of the distribution can cause impact ionization, generating electron–hole pairs. In n-channel FETs, the generated holes are swept into the bulk, while the electrons are injected into the oxide, leading to long-term reliability degradation. In addition to hot-carrier degradation, the substrate current caused by hole injection into the substrate produces an effect similar to that of a small positive substrate bias, resulting in a reduction of the threshold voltage.

2.5 MOSFET DEVICE MODELS

2.5.1 The Level 1 Model

The Level 1 SPICE model, also known as the Shichman–Hodges model [20], uses Equations 2.70 and 2.71 to arrive at the well-known drain current equation for the linear region:

$$I_{ds} = \frac{\mu C_{ox} W}{L}\left[\left(V_{gs} - V_{th}\right)V_{ds} - \frac{V_{ds}^2}{2}\right] \tag{2.74}$$

Note that, here, W and L represent the effective width and length, respectively, that is, the reduced dimensions after subtracting out the overlap of the gate with the diffusion regions caused by lateral diffusion. The saturation voltage is given by

$$V_{dsat} = V_{gs} - V_{th} \tag{2.75}$$

so that, in saturation, the drain current becomes (from Equations 2.74 and 2.75)

$$I_{ds} = \frac{\mu C_{ox} W}{2L}\left(V_{gs} - V_{th}\right)^2 \tag{2.76}$$

Channel length modulation is also accounted for using

$$I_{ds} = I_{dsat}(1+\lambda V_{ds}) \tag{2.77}$$

where I_{dsat} is as given by Equation 2.76, and λ is the channel length modulation parameter. Drain and source resistances are included in the model, and there is also a parameter to represent surface state density, which models the contribution of interface states at the oxide–silicon boundary to the total charge density.

The FET capacitances are calculated from the node charges. Computation of the charge is model-specific, in that it depends on the particular model, whereas computation of the capacitance is circuit simulator specific. Computation of the node charges begins with charge neutrality, by which

$$Q_G + Q_{INV} + Q_{DEP} = 0 \tag{2.78}$$

where Q_G is the gate charge, Q_{INV} is the inversion charge, and Q_{DEP} is the depletion charge. The depletion charge is given by

$$Q_{DEP} = WLC_{ox}\gamma(2\varphi_f + V_{sb})^{1/2} \tag{2.79}$$

The inversion charge is given by

$$Q_{INV}(y) = -\ WLC_{ox}\ (V_{gs} - V_{th} - V(y)) \tag{2.80}$$

so that the total gate charge becomes

$$Q_G = -W\int_0^L Q_{inv}(y)dy - Q_{DEP} = \frac{\mu C_{ox}^2 W^2}{I_{ds}}\int_0^L \left(V_{gs} - V_{th} - V\right)^2 dV - Q_{DEP} \tag{2.81}$$

Using the appropriate current equations, the charge in different regions of operation can be calculated. In SPICE, these charge expressions are used to calculate the capacitances (the so-called C–V portion of the model) using the Meyer model [21].

2.5.2 The Level 2 and Level 3 Models

The only small-geometry effect included in the Level 1 model is channel length modulation (by means of the parameter λ). The Level 2 model follows a similar modeling approach to Level 1, but adds corrections to include more small-geometry effects. The variation of charge along the length of the channel is accommodated, leading to a more accurate expression for the drain current. Carrier velocity saturation and the consequent reduction in V_{dsat} are modeled, as is mobility degradation by the vertical electric field. However, the mathematical implementation of the model is complicated, and the model suffers from a severe discontinuity in the first derivative at the transition from linear region to saturation, leading to convergence problems during simulation. The Level 3 model aims to overcome these issues. It is semiempirical and is a simplified version of Level 2. It is computationally more efficient, and its robustness made it very popular for digital IC design. Some additional effects, such as, DIBL and mobility degradation due to the lateral field, are included. However, it is not very scalable and binning is almost always required. This is mainly because the Level 3 model treats short and narrow channel effects as corrections to the basic long, wide channel device equations. Thus, for accurate modeling of different device dimension configurations, that is, different regions in the device geometry space, different correction factors would be required. At least four subspaces can immediately be recognized: long, wide devices; short, wide devices; long, narrow devices; and short, narrow devices. This results in a "binned" model, with the appropriate correction factors used according to the device geometry. Level 1, 2 and 3 SPICE models are considered first-generation MOSFET models.

2.5.3 BSIM

BSIM (Berkeley short-channel IGFET model) [22] is usually considered the first second-generation MOSFET model. It represents a significant change in modeling approach from the first-generation models, in that, instead of expending enormous effort to develop accurate analytical model equations without regard for complexity, the focus is on using empirical formulations with good mathematical conditioning, to enable faster and more robust circuit simulation. One consequence is that a great many empirical parameters had to be introduced to achieve this objective, increasing the effort in extracting these parameters, and reducing their direct correspondence to physical device phenomena. The equations are mathematically "conditioned" to improve their convergence and efficiency in circuit simulation, and a new parameter structure is introduced to describe the geometry dependence of parameters. Many parameters that are included in the model to satisfy these objectives do not bear clear physical meaning. The quality of the model becomes heavily dependent on parameter extraction.

BSIM models mobility similar to the first-generation models. Mobility is described as a basic mobility μ_0, which is modified by the effect of vertical and lateral electric fields. The difference here is that μ_0 is a composite parameter derived from other model parameters. The use of composite parameters, where model parameters are constructed from other model parameters, is an important feature of second-generation MOSFET models. Using this technique, μ_o is modeled to be dependent on V_{bs} and V_{ds} (this is achieved by using different expressions for μ_o at V_{ds}=0 and V_{DD}, and using quadratic interpolation between these two points). The actual mobility μ showing dependence on the vertical field, is modeled as

$$\mu = \frac{\mu_0}{1 + U_0 \left(V_{gs} - V_{th} \right)} \tag{2.82}$$

U_0 itself is a composite parameter, and includes substrate bias dependence. The mobility model also includes the effect of the lateral field, but this is a consequence of the drain current and saturation voltage models.

The threshold voltage V_{th} is modeled as

$$V_{th} = V_{fb} + 2\varphi_f + \gamma\,(2\varphi_f + V_{sb})^{1/2} - K_2\,(2\varphi_f + V_{sb}) - \eta V_{ds} \tag{2.83}$$

where K_2 is the drain/source depletion charge-sharing coefficient and η is a DIBL related parameter.

The BSIM expression for the drain current in the linear region may be given as,

$$I_{ds} = \frac{\mu C_{ox} W}{L} \left[\left(V_{gs} - V_{th} \right) V_{ds} - \frac{a V_{ds}^2}{2} \right] \tag{2.84}$$

While this might appear to be a simple expression, a great deal of complexity is hidden in the modeling of the various parameters like μ, V_{th}, and a. BSIM models the decrease in effective saturation voltage in short-channel devices due to carrier velocity saturation (Note that the attempt to include this effect started with the Level 2 model; however, the resultant discontinuity in the first derivative of the current at V_{dsat} led to convergence issues). BSIM reevaluates the inversion charge Q_{inv} as

$$Q_{inv} = C_{ox}\,(V_{gs} - V_{th} - a\varphi_n) \tag{2.85}$$

where the parameter a is the same as the one in Equation 2.84, and φ_n is the electron quasi-Fermi level. From this, the saturation drain current can be derived to be of the form

$$I_{ds} = \frac{\mu C_{ox} W}{2L} \left(\frac{1}{1 + \frac{U_1}{L}(V_{gs} - V_{th})} \right) (V_{gs} - V_{th})^2 \tag{2.86}$$

where U_1 is defined as

$$U_1 = \frac{\mu}{v_{sat}} \tag{2.87}$$

where v_{sat} is the carrier saturation velocity. Under short-channel conditions (i.e., small L), this expression reduces to

$$I_{dsat} = v_{sat} C_{ox} W (V_{gs} - V_{th}) \tag{2.88}$$

Notice that the original quadratic dependence of the channel current on (V_{gs}–V_{th}) has changed to a linear dependence. Also, the current is no longer dependent on the channel length. The behavior of drain saturation current in short-channel devices is dominated by velocity saturation. Finally, V_{dsat} is modeled in BSIM as

$$V_{dsat} = \frac{1}{a} \left(V_{gs} - V_{th} - \frac{I_{dsat}}{a C_{ox} v_{sat}} \right) \tag{2.89}$$

The charge model in BSIM is based on that developed by Yang et al. [23], and is computed for four regions of operation--accumulation, subthreshold, linear and saturation--with care taken to ensure continuity at the region boundaries. In accumulation, the inversion charge is zero and the gate charge is mirrored in the substrate. In strong inversion, on the other hand, the inversion charge is not negligible. The presence of significant inversion charge makes it necessary to have a partitioning scheme for the inversion charge between the drain and the source. BSIM provides three choices for this partitioning between drain and source: 2:3, 1:1, and 0:1. The node charge equations for the linear and saturation regions are given by Equations 2.90–2.95. In the below equations, Q_G is the total gate charge, Q_{DEP} is the total depletion charge, and Q_{INV} is the total inversion charge. α_x is a parameter describing short-channel effects, and is defined as in Equation 2.91.

Node charge equations for the linear region:

$$Q_G = WLC_{ox} \left(V_{gs} - V_{fb} - 2\varphi_f - \frac{V_{ds}}{2} + \frac{\alpha_x V_{ds}^2}{12\left(V_{gs} - V_{th} - \frac{\alpha_x V_{ds}}{2} \right)} \right) \tag{2.90}$$

$$\alpha_x = a\left(1+\frac{U_1}{L}\left(V_{gs}-V_{th}\right)\right) \tag{2.91}$$

$$Q_{DEP} = WLC_{ox}\left(V_{fb}+2\varphi_f-V_{th}+\frac{(1-\alpha_x)V_{ds}}{2}-\frac{(1-\alpha_x)\alpha_x V_{ds}^2}{12\left(V_{gs}-V_{th}-\frac{\alpha_x V_{ds}}{2}\right)}\right) \tag{2.92}$$

Node charge equations for the saturation region:

$$Q_G = WLC_{ox}\left(V_{gs}-V_{fb}-2\varphi_f-\frac{V_{gs}-V_{th}}{3\alpha_x}\right) \tag{2.93}$$

$$Q_{DEP} = WLC_{ox}\left(V_{fb}+2\varphi_f-V_{th}+\frac{(1-\alpha_x)(V_{gs}-V_{th})}{3\alpha_x}\right) \tag{2.94}$$

$$Q_{INV} = -\frac{2}{3}WLC_{ox}\left(V_{gs}-V_{th}\right) \tag{2.95}$$

2.5.4 The BSIM2 and HSPICE Level 28 Models

BSIM, in spite of being a major improvement over earlier models, has some shortcomings. The use of polynomial equations instead of physical expressions often leads to convergence problems and a negative output conductance. It is also oriented more toward digital circuit design, and not very analog-design-friendly. Two models were proposed to overcome these deficiencies. One is the HSPICE Level 28 model, which uses a modified description in the linear–saturation transition region, and includes provision for binning, to improve the suitability of the original BSIM model for analog design. The low-field mobility is reformulated to remove its quadratic dependence on V_{ds}. A transition region is defined around V_{dsat}, and a smoothing function is used for the current in this region. The other model proposed to mitigate the shortcomings of BSIM is the BSIM2 model. The mobility and drain current models are modified from BSIM. A new subthreshold current model gives better accuracy and convergence. A transition region is defined around the weak inversion–strong inversion boundary (this is also done in HSPICE Level 28). Modeling of the output conductance is improved by adding or improving equations to model drain-induced barrier lowering, channel length modulation, and hot-carrier-induced changes in the bulk potential. This contributes to making BSIM2 suitable for analog circuit design. BSIM2 improves on BSIM both in terms of model accuracy and convergence properties. However, transistor operation is still divided into different regions of operation, and discontinuity is present in the first derivative in the I–V and C–V characteristics, which can cause numerical issues in simulation.

2.5.5 BSIM3

The second-generation MOSFET models are very empirical and complex, and contain a large number of equations and parameters. While the original objective of making the models more attuned to circuit simulation using empiricism was well intentioned, evolution of device technology meant that more parameters and equations were needed in second-generation models, to accurately characterize device behavior. This led to the model becoming large and unwieldy, and such models can be slow and inefficient during circuit simulation. Parameter extraction can be complex and tedious, requiring several test structures. To overcome these shortcomings, third-generation MOSFET models return to a simpler model structure with a reduced number of parameters. This is made possible by making them less empirical and more physically based. The main objective of third-generation models is to reduce the parameter set by building a better connection between the parameters and the physics of the device, while retaining the mathematical fitness of the model equations. BSIM3 is the most popular third-generation MOSFET model. BSIM3 has evolved through three versions. BSIM3v1 exhibited some severe mathematical problems, which are corrected in BSIM3v2. BSIM3v3 (which itself has grown into multiple versions) is the most popular version of the model. It introduced changes to ensure continuity in the equations, along with many empirical fitting expressions and additional parameters to improve accuracy. A departure from second-generation models is that the model structure is not designed to accommodate binning, and the model equations are taken to be valid for all device geometries. Smoothing functions are extensively used to improve mathematical behavior in the transition regions between adjacent regions of operation. It also unifies the descriptions in different operating regions into single expressions for mobility, channel charge density, drain current, and V_{dsat}. This eliminates the discontinuities in the I–V, C–V, and conductance characteristics (and those of the derivatives), and enables better convergence and efficiency during circuit simulation. For instance, the general form of the drain current I_{ds} may be given as

$$I_{ds} = \frac{I_{ds0}}{1 + \dfrac{R_{ds} I_{ds0}}{V_{ds,eff}}} \left(1 + \frac{V_{ds} V_{ds,eff}}{V_A}\right) \tag{2.96}$$

$$I_{ds0} = \frac{\mu_{eff} W_{eff} C_{ox} V_{gst,eff} V_{ds,eff}}{\left(1 + \dfrac{V_{ds,eff}}{\varepsilon_{sat} L_{eff}}\right)} \left(1 - \frac{A_{bulk} V_{ds,eff}}{2V_{gst,eff} + 4V_T}\right) \tag{2.97}$$

where $V_{ds,eff}$ is the effective drain-source voltage, $V_{gst,eff}$ is the effective saturation voltage, and A_{bulk} is the bulk charge coefficient. Detailed descriptions of BSIM3 can be found in references [24–26].

It is interesting to note that while the original objective of BSIM3 was to achieve a relatively simple model with reduced empirical content and a small parameter set, as it evolved, many empirical expressions were introduced, with the attendant large number of parameters. In terms of RF IC design and simulation, BSIM3 has some serious shortcomings. For instance, no gate resistance is included, and the resistance in the substrate is neglected. This results in inaccurate prediction of circuit characteristics at high frequencies. Chapter 4 describes modifications to the standard BSIM3v3 model to correct these deficiencies and enable accurate modeling of high-frequency device behavior.

2.5.6 MOS Model 9 and MOS Model 11

MOS Model 9 and MOS Model 11, developed by Philips, are the most popular non-Berkeley MOSFET models. MOS Model 9 uses the third-generation approach of utilizing smoothing functions to achieve continuity across different regions of device operation. However, unlike BSIM3, MOS Model 9 retains the second-generation approach of describing the geometry dependence of model parameters, albeit with extensive modifications to achieve better accuracy. The model is geared toward circuit simulation, and uses a relatively small number of equations and model parameters. MOS Model 11 is the successor to MOS Model 9, and features bias-dependent series resistance and bias-dependent overlap capacitance models, and accurate modeling of gate leakage current, gate-induced drain leakage, gate depletion and quantum mechanical effects. It is a surface-potential based model and utilizes smoothing functions to achieve better convergence in simulation. More details regarding MOS Models 9 and 11 may be found on their websites [27,28].

2.5.7 BSIM4

BSIM4 features many improvements over BSIM3 in the drain current model, noise modeling, modeling of extrinsic parasitics, and includes models for gate-induced drain leakage current, gate and substrate resistances, and drain-induced gate noise. It accounts for finite inversion layer thickness and improves the accuracy of modeling of the C–V characteristics. It is the preferred model for CMOS technology generations from 0.18 μm onward. The drain current formulation of BSIM4 is similar to that of BSIM3 (i.e. as given by Equations 2.96 and 2.97); however the oxide capacitance C_{ox} of Equation 2.97 is replaced by effective oxide capacitance $C_{ox,eff}$ which accounts for the difference between the physical and electrical oxide thicknesses. BSIM4 is designed to overcome some of the deficiencies of BSIM3 with respect to high-frequency characteristics. It provides the option of incorporating gate and substrate resistances using a settable model parameter. It allows significant layout-related information, such as the number of gate fingers, number of contacts per finger, and drain/source configuration (shared or isolated) to be included in the transistor model statement, which enables better scaling of the model for RF applications. BSIM4 fixes asymmetry problems encountered with BSIM3 using a dynamic reference approach. It features a holistic thermal noise model that recognizes the existence of induced gate noise, which is important in preventing nonphysical noise figure prediction during simulation. The problem of incompatibility of noise simulation in SPICE, which uses superposition of the effect of different noise sources, with the actual physical fact that channel current noise and drain-induced noise are significantly correlated, is fortuitously circumvented by transforming the gate noise source to the source. It turns out that the correlation between the drain noise current and the transformed noise current at the source is negligible, enabling SPICE to use superposition without losing accuracy. More details regarding BSIM4 can be found in references [24,25].

REFERENCES

1. J. J. Ebers and J. L. Moll, Large-signal behavior of junction transistors, *Proc. IRE*, **42**: 1761–1772 (Dec. 1954).
2. I. E. Getreu, *Modeling the Bipolar Transistor*, Elsevier Scientific Publishing Company, New York; 1978.
3. H. K. Gummel and H. C. Poon, An integral charge control model of bipolar transistors, *Bell Syst. Tech. J.*, **49**: 827–852 (May 1970).
4. C. C. McAndrew, J. A. Seitchik, D. F. Bowers, M. Dunn, M. Foisy, I. Getreu, M. McSwain, S. Moinan, J. Parker, D. J. Roulston, M. Schroter, P. van Wijnen, and L. F. Wagner, VBIC95, the vertical bipolar inter-company model, *IEEE J. Solid-State Circuits*, **31**(10): 1476–1483 (Oct. 1996).
5. W. J. Kloosterman and H. C. de Graaf, Avalanche multiplication in a compact bipolar transistor model for circuit simulation, *IEEE Trans. Electron Devices*, **36**(7): 1376–1380 (July 1989).
6. G. M. Kull, L. W. Nagel, S.-W. Lee, P. Lloyd, E. J. Prendergast, and H. K. Dirks, A unified circuit model for bipolar transistors including quasi-saturation effects, *IEEE Trans. Electron Devices*, **32**(6): 1103–1113 (June 1985).
7. H. C. de Graff, W. J. Kloosterman, J. A. M. Geelan, and M. C. A. M. Koolen, Experience with the new compact MEXTRAM model for bipolar transistors, *Proc. Bipolar Circuits and Technology Meeting*, 1989, pp. 246–249.

8. H. C. de Graff and W. J. Kloosterman, New formulation of the current and charge relations in bipolar transistor modeling for CACD purposes, *IEEE Trans. Electron Devices*, **32**(11): 2415–2419 (1985).
9. H. C. de Graff, W. J. Kloosterman, J. C. J. Paasschens, R. van der Toorn, R. J. Havens, and J. A. M. Geelan, *MEXTRAM documentation* (online), available at http://www.nxp.com/Philips_Models/bipolar/mextram.
10. H. Stübing and H.-M. Rein, A compact physical large-signal model for high-speed bipolar transistors at high-current densities, Part I: One-dimensional model, *IEEE Trans. Electron Devices*, **34**: 1741–1751 (1987).
11. H.-M. Rein and M. Schröter, A compact physical large-signal model for high-speed bipolar transistors at high-current densities, Part II: Two-dimensional model and experimental results, *IEEE Trans. Electron Devices*, **34**: 1752–1761 (1987).
12. H.-M. Rein, M. Schröter, A. Koldehoff, and K. Wörner, A semi-physical bipolar transistor model for the design of very high-frequency analog ICs, *Proc. IEEE Bipolar Circuits and Technology Meeting*, 1992, pp. 217–220.
13. M. Schröter, M. Friedrich, H.-M. Rein, A generalized integral charge-control relation and its application to compact models for silicon based HBTs, *IEEE Trans. Electron Devices*, **40**(11): 2036–2046 (Nov. 1993).
14. D. Berger, D. Cell, M. Schröter, M. Malorny, T. Zimmer, and B. Ardouin, HICUM parameter extraction methodology for a single transistor geometry, *Proc. IEEE Bipolar/BiCMOS Circuits and Technology Meeting*, 2002, pp. 116–119.
15. M. Schröter and A. Chakravorty (2005), HICUM, a geometry scalable physics-based compact bipolar transistor model, User's Manual HICUM/Level2 version 2.2 (online), available at http://www.iee.et.tu-dresden.de/iee/eb/hic_new/hic_doc.html.
16. M. Schröter, High-frequency circuit design oriented compact bipolar transistor modeling with HICUM, *IEICE Trans. Electron.*, **E88-C**(6): 1098–1113 (June 2005).
17. M. Schröter, S. Lehmann, H. Jiang, and S. Komarow, HICUM/Level0 – A simplified compact transistor model," *Proc. IEEE Bipolar/BiCMOS Circuits and Technology Meeting*, 2002, pp. 112–115.
18. M. Schröter, S. Lehmann, S. Fregonese, and T. Zimmer, A computationally efficient physics-based compact bipolar transistor model for circuit design – Part I: Model formulation, *IEEE Trans. Electron Devices*, **53**(2): 279–286 (Feb. 2006).
19. S. Fregonese, S. Lehmann, T. Zimmer, M. Schröter, D. Celi, B. Ardouin, H. Beckrich, P. Brenner, and W. Kraus, A computationally efficient physics-based compact bipolar transistor model for circuit design – Part II: Parameter extraction and experimental results, *IEEE Trans. Electron Devices*, **53**(2): 287–295 (Feb. 2006).
20. H. Schichman and D. A. Hodges, Modeling and simulation of insulated-gate field-effect transistor switching circuits, *IEEE J. Solid-State Circuits*, **3**(3): 285–289 (Sept. 1968).
21. J. Meyer, MOS models and circuit simulation, *RCA Review*, **32**: 42–63 (1971).
22. B. J. Sheu, D. L. Scharfetter, P.-K. Ko, and M.-C. Jeng, BSIM: Berkeley Short Channel IGFET Model for MOS transistors, *IEEE J. Solid-State Circuits*, **22**(4): 558–566 (Aug. 1987).
23. P. Yang, B. D. Epler, and P. K. Chatterjee, An investigation of the charge conservation problem for MOSFET circuit simulation, *IEEE J. Solid-State Circuits*, **18**(1): 128–138 (Feb. 1983).
24. BSIM homepage (online), available at http://www- device.eecs.berkeley.edu/~bsim3/.
25. W. Liu, *MOSFET Models for SPICE Simulation, Including BSIM3v3 and BSIM4*, Wiley, New York, 2001.

26. D. Foty, *MOSFET Modeling with SPICE: Principles and Practice*, Prentice-Hall, Englewood Cliffs, NJ, 1997.
27. MOS Model 9 homepage (online), available at http://www.nxp.com/Philips_Models/mos_models/model9.
28. MOS Model 11 homepage (online), available at http://www.nxp.com/Philips_Models/mos_models/model11.

3

EMPIRICAL MODELING OF BIPOLAR DEVICES

3.1 Introduction

The bipolar IC design community has largely relied on the SPICE Gummel–Poon (SGP) model for nearly three decades, till the recent introduction of other models like VBIC, MEXTRAM, and HICUM. This is in contrast to the tremendous growth in the number of MOSFET models, with over 50 versions available in HSPICE alone. The formulation of the Gummel–Poon model accounts for the major physical mechanisms that control BJT behavior. However, the approximations that underlie the SGP model ignore effects that are important for accurate modeling of today's BJTs. The simplified SGP Early effect model is insufficient to accurately model the output conductance, and collector resistance modulation is ignored. Avalanche multiplication and self-heating are not modeled, and the modeling of high-frequency and temperature effects is inadequate. More recent physics-based models, such as VBIC and MEXTRAM, overcome many of these deficiencies. While these models address the issues in modeling the BJT, the case of the HBT is more complicated.

3.1.1 Modeling the HBT versus the BJT

Conventional homojunction BJT models, such as the Ebers–Moll or the Gummel–Poon model, are widely used for HBT-based circuit design. However, the validity of these models for the HBT is limited because of the several significant differences in the electrical behavior of HBTs and BJTs. For instance, typically, a constant-current-gain (midcurrent) region is present in the BJT forward Gummel characteristic, where the base current is due largely to recombination in the neutral base region, and as a result, the ideality factors of the base and collector currents are approximately the same. However, this is generally absent in an HBT forward Gummel characteristic. This is because, in the HBT, several recombination current components, namely, the recombination current in the emitter–base space charge region (SCR), the recombination current at the emitter–base heterojunction, the recombination current at the edge of the emitter–base SCR, and the recombination current in the neutral base region, contribute to the base current, over the entire forward active region [1]. Further, unlike a BJT, the Early effect is usually negligible in an HBT, as the high base doping results in a very high Early voltage. On the other hand, self-heating is severe in a GaAs-based HBT compared to a BJT, and any HBT model must account for this effect.

Many HBT-specific, as well as Ebers–Moll/Gummel–Poon/VBIC models modified for the HBT, have been proposed [2–11]. However, because of the presence of strong self-heating, and the absence of a well-defined constant current–gain region in the forward–Gummel characteristics of a typical GaAs-based HBT, parameter extraction using the traditional Gummel-plot-based extraction procedure is usually found to result in an inadequate fit to the measured DC characteristics. An additional optimization step is generally found to be necessary to obtain good correspondence between measured and modeled characteristics.

3.1.2 Parameter Extraction

A device model is only as good as the accompanying parameter extraction technique. An exhaustive set of model equations is of little use if the model parameters cannot be extracted accurately and reliably. The extraction algorithm must be viewed as an integral part of the transistor model. As device models evolve to a higher level of sophistication, the complexity of parameter extraction increases manifold. The classical method of extracting bipolar DC model parameters involves extraction of the diode-like exponential current source model parameters from the Gummel plots, and the addition of thermal resistance, which is extracted separately, usually from pulsed I–V measurements [7,12]. This is the so-called direct extraction approach [13]. In a direct extraction procedure, extraction is performed using analytical expressions for all model parameters, making the extraction fast and well-defined. In bipolar device technologies with significant self-heating, in practice, the extraction of parameters from the Gummel plots has to be done under pulsed conditions, or at biases where self-heating is insignificant. Thus, the extraction process becomes quite cumbersome and tedious, and often requires some optimization after extraction, to obtain an acceptable fit to the measured DC I–V characteristics, thereby no longer retaining the desirable direct extraction feature. On the other hand, extraction procedures that use brute-force optimization suffer from numerous problems, such as, nonuniqueness of the extracted parameter values, trapping at local minima, and requirement of substantial compute time, defects that the direct extraction approach lacks. In the direct extraction approach, it is possible to reduce the number of measurement points needed to a minimum, approaching the number of model parameters to be extracted. However, sensitivity to noise can then increase considerably. By using more measurement data than the minimum required, and using traditional least-squares techniques, analytical expressions can still be used, making this an attractive combination of "pure" direct extraction and optimization (or error minimization). The exponential-based analytical bipolar current source equations, in particular, lend themselves to an elegant formulation that can utilize a direct parameter extraction algorithm using the least squares technique, which inherently incorporates optimization, that is, the minimization of average least-squared error between measured and modeled data. The modeling methodology described in this chapter represents this happy marriage between direct extraction and optimization, which yields curve-fitted parameter values for a physics-based analytical model formulation. Given this parameter extraction philosophy, it remains to cast the analytical device equations in a form that can utilize this approach. This will be shown to involve making intelligent approximations to the various terms of the terminal current equations that represent different device physical phenomena, to make these terms compatible with this model formulation philosophy, yet be accurate enough to represent the respective phenomena adequately.

3.1.3 Motivation for an Empirical Bipolar Device Model

Bipolar transistor technologies are a favorite choice for RF IC implementation. AlGaAs/GaAs, and more recently, InGaP/GaAs HBTs, are the technologies of choice for RF power amplifier design, especially for handsets. SiGe HBT has also emerged as a viable choice for power amplifier design. RF front ends are often designed in Si BJT or SiGe BiCMOS technology. The motivation for developing bipolar equivalent-circuit models that are accurate for high-frequency and large-signal design, is therefore, undoubtedly immense. Yet, until recently, the outdated and inadequate SPICE Gummel–Poon (SGP) model was virtually the only model available for bipolar transistors. In the last few years, however, there has been a surge in activity to fulfill the urgent need for an accurate bipolar large-signal model. As a result, new models, such as the VBIC, MEXTRAM, and HICUM models, have been proposed. These were developed mainly for the BJT, and are all physics-based models, continuing in the tradition of physics-based bipolar modeling. These models do not inherently model the HBT well, as explained previously. There has also been a tremendous amount of effort expended to develop HBT-specific large-signal models. However, we still do not have a unified, accurate model that has been accepted as the standard model for the HBT. The reason for this is that the device physics of an HBT is very complicated, with several phenomena that have to be modeled, and parameters describing these phenomena have to be extracted. Further, especially in GaAs-based HBTs, thermal conductivity is poor, and the self-heating effect is severe. All these factors make it virtually impossible to have a completely physics-based model description. Most models, although based on a physical foundation, end up with several empirical coefficients that are difficult to extract. In the context of power amplifier design, these difficulties increase several fold, because the complex thermal interactions between the different fingers in a multifinger power cell, or those between unit cells in a large power transistor that comprises several of these unit cells, have to be accounted for in the model. These thermal interactions are highly dependent on the particular layout of the power transistor, are highly nonlinear, and manifest themselves as unpredictable effects on the scaling of the model parameters. This necessitates either the use of arbitrary empirical scaling rules, or a laborious thermodynamic analysis of heat flow in the power transistor structure to create a complex electrothermal model. Not only is the latter daunting and time-consuming to develop; the sheer complexity of the resulting model often makes it infeasible to implement in a circuit simulator. These arguments serve to nullify the primary limitation of an empirical model compared to a physics-based model, from a circuit design perspective; namely, an empirical model is valid only for the device for which it was extracted, whereas a physics-based model could be scaled to predict the characteristics of various device geometries. It is clear from the discussion above that the scaling ability of a purely physics-based model is limited; indeed, it is quite inadequate for the extent required in power amplifier applications. In this context, given that physics-based models, especially of the HBT, often end up with a significant empirical content and any attempt at scaling these models requires an empirical approach, and given the arguments made in the previous section on parameter extraction, the rationale for adopting a completely empirical approach to develop a bipolar large-signal model is self-evident. Such an empirical approach becomes even more attractive if it is accompanied by a simple and fast extraction algorithm that eliminates explicit optimization effort.

This chapter describes a novel empirical bipolar large-signal model, including self-heating effects. The model is focused on the forward active region of operation, since this is the normal region of operation of most practical RF circuits. The model accounts for the inherent temperature dependence of the device characteristics due to both ambient-temperature variation and self-heating. The model is accompanied by a simple extraction process, which requires only DC current–voltage (I–V), and multi-bias-point small-signal S-parameter measurements. All the current-source model parameters, including the self-heating parameters, are directly extracted from measured forward I–V data at different ambient temperatures. The extraction procedure is fast, accurate, and inherently minimizes the average squared error between measured and modeled data, thereby eliminating the need for further optimization following parameter extraction. The bias-dependent nonlinear intrinsic elements are empirically extracted from small-signal S-parameter measurements at different bias conditions and temperatures. The distributed base-collector capacitance and base resistance are extracted from measured S-parameters using a new technique. The model is unified; that is, it predicts the DC, and high-frequency small-signal and large-signal characteristics of the device accurately. The model is focused on the GaAs HBT (it is validated for an InGaP/GaAs HBT device), but the applicability of the modeling technique for the SiGe HBT and the Si BJT is also demonstrated.

3.1.4 Physics-Based and Empirical Models

At this juncture, it is relevant to distinguish between "physics-based" models and purely physical models. Physical models are based on a description of the physical processes that characterize the transistor. This requires a description of the carrier transport physics and the associated geometric and material properties of the transistor. Such models require enormous computational power, and are usually used only in device design. In physics-based equivalent-circuit models (often called "compact models"), to which category models like SGP, VBIC, and MEXTRAM belong, the element values are obtained from expressions that phenomenologically relate the circuit elements to device physics. These models utilize the physical data of the transistor, and can be useful in relating transistor performance to the geometry and physics of the device, but they often require a detailed set of measurements and empirical fitting factors. These models require a well-characterized fabrication process, where the physical data required for the development of the model are available.

Empirical models also typically utilize an equivalent-circuit formulation inspired by the physics of the device, but they do not require knowledge of the device physical data. The parameters of the model are derived by curve fitting to measured data. Traditionally, MOSFET models incorporate a higher degree of empiricism than bipolar models, because the nature of MOSFET device physics precludes a completely analytical description of many of the physical phenomena in the device. A similar situation exists in the case of GaAs-based HBTs, where an empirical approach is generally used to extract the intrinsic bias-dependent nonlinear elements, due to difficulties in arriving at an analytical solution for these elements (e.g., the base–collector junction capacitance) because of self-heating, and the nonlinear field dependence of the electron velocity.

3.1.5 Compatibility between Large- and Small-Signal Models

An important consideration in the development of a unified large-signal model is the compatibility between large-signal and small-signal models. Most bipolar small-signal models adopt a hybrid T topology [14–16], since it directly relates to the physics of the device. However, the popular Gummel–Poon-like large-signal model topology reduces to a hybrid π rather than a T topology under small-signal conditions. In order to obtain compatibility between the large and small-signal models, the hybrid π topology has been adopted for the small-signal model, in the model described in this chapter. The bias-dependent nonlinear intrinsic elements are extracted from small-signal S-parameter measurements, using this topology. Also, the π model parameters are known to show frequency dependence, especially at higher frequencies [17]. The model described in this chapter uses a direct extraction procedure for the intrinsic capacitances, which inherently minimizes the average squared error between modeled and measured data over the entire frequency range of interest.

3.2 MODEL CONSTRUCTION AND PARAMETER EXTRACTION

The construction of the empirical large-signal bipolar model will first be elaborated for one bipolar device, for convenience. The GaAs (InGaP/GaAs) HBT is chosen for this purpose. The modifications required in the case of the SiGe HBT and Si BJT will be discussed subsequently. It will be shown that the same modeling and parameter extraction philosophy works for all the devices, with refinements to be applied in each case according to the physics of the device.

Figure 3.1 shows the proposed large-signal equivalent-circuit model of the InGaP/GaAs HBT, and Figure 3.2 shows the corresponding small-signal model. The large-signal model is composed of diode-like current sources; bias-dependent capacitances, including a distributed base–collector capacitance; distributed base resistance; a thermal subcircuit to model self-heating, which dynamically modifies pertinent electrical parameters; and the access parasitic elements, which model pad capacitances, and access resistances and inductances.

3.2.1 Curent Source Model

The current source model consists principally of two diode-like current sources: I_{BE} and I_{CC}. The I_{BC} diode of the traditional Gummel–Poon model is omitted, as it is significant only in inverse region operation and strong saturation. Given that I_{SC} represents the preexponential factor, V_{bei} is the internal base–emitter voltage (after deembedding the total base resistance, R_{btot}, and emitter resistance, R_e), I_B is the base current, n_C is the ideality factor, and R_{bc} is an empirical resistance used to adjust the high-current region current gain [3,18], the formulation of the I_{CC} current source is described with a conventional diode-like equation:

$$I_{CC} = I_{SC}\left[\exp\left(\frac{V_{bei} - I_B R_{bc}}{n_C V_T}\right) - 1\right] \tag{3.1}$$

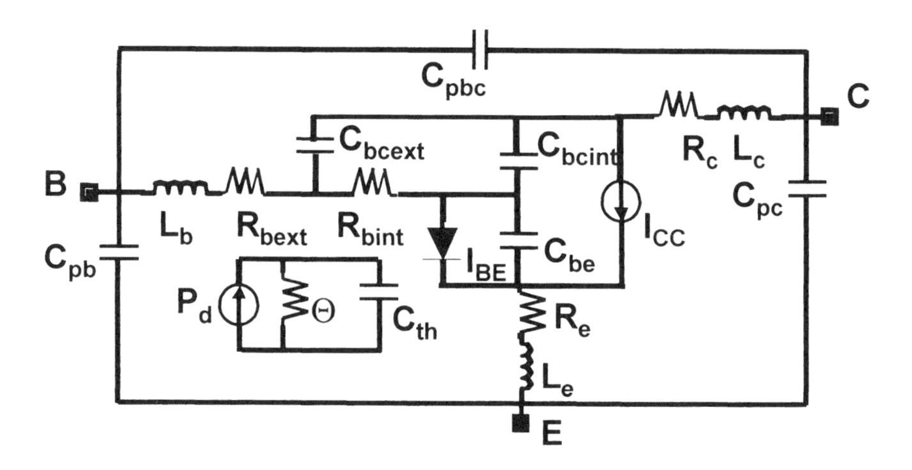

Figure 3.1. InGaP/GaAs HBT large-signal equivalent circuit.

The temperature dependence of the ideality factor n_C is modeled as

$$n_C = n_{C0} + \delta \Delta T \tag{3.2}$$

where δ represents the thermal dependence of the ideality factor n_C, whose value is n_{C0} at 0 K, and ΔT, the effective junction temperature, is the sum of the ambient temperature T_A and the increment in temperature due to self-heating, given by ΘP_d, where Θ is the thermal resistance and P_d is the power dissipation. From Equation 3.1, since δ is small, Equation 3.3 can be derived to model I_{CC} at a given ambient temperature [19,20]:

$$I_{CC} = I_{SC} \cdot \exp(-a\Theta P_d) \exp\left[\frac{V_{bei}}{n_C V_T}\left(1 - \frac{\delta \Theta P_d}{n_C}\right) - \frac{I_B R_{bc}}{n_C V_T}\right] \tag{3.3}$$

where the empirical factor $f_{SH} = \exp(-a\Theta P_d)$, with a fitting parameter a, is introduced to model self-heating. The dependence of Θ on the junction temperature is modeled as $\Theta = \Theta_0(1+bP_d)$.

To overcome the limitation in modeling the saturation region caused by omission of the I_{BC} diode in the model, an empirical factor, $f_{sat} = \exp(a_{sat} \cdot V_{bcirev})$, is introduced. Here $V_{bcirev} = V_{bci}$ when $V_{bci} < 0$, and 0 otherwise, so that this term contributes only in the saturation regime.

An empirical term, $f_I = \exp(a_I I_C)$, is introduced to model the current gain at high bias current, analogous to the knee currents of the standard Gummel–Poon model. It also contributes to modeling the current dependence of the empirical resistance R_{bc}. Another factor, $f_V = \exp(a_I V_{cei})$, empirically facilitates the modeling of the Kirk effect and heterojunction barrier effect, which helps improve model accuracy at lower collector–emitter voltages and high bias current density, analogous to the empirical functions proposed by Sinnesbichler and Olbrich [21]. The resulting model for the I_{CC} current source at a particular ambient temperature is given by

$$I_{CC} = I_{SC0} \cdot f_{SH} \cdot f_{I} \cdot f_{V} \cdot f_{sat} \cdot \exp\left[\frac{V_{bei}}{n_C V_T}\left(1 - \frac{\delta \Theta P_d}{n_C}\right) - \frac{I_B R_{bc}}{n_C V_T}\right] \tag{3.4}$$

All the terms in this equation vary with ambient temperature, including the pre-exponential factor I_{SC0}, since some effects are partially absorbed into this term.

The base–emitter diode I_{BE} is modeled similarly to I_{CC}. Since the physical mechanism that controls the base current is distinct from the mechanism that controls the collector current, the two currents are not linked by a phenomenological parameter (i.e., β). Also, since the thermal resistance is physically the same for both current sources, the self-heating factor for the base current source is defined as $f_{SHB} = \exp\{-a_{SHB}\Theta_0(1+bP_d)P_d\}$, where Θ_0 and b are the same for both the base and collector current sources. The model for the I_{BE} diode at a particular ambient temperature is therefore given by

$$I_{BE} = I_{SB0} \cdot f_{SHB} \cdot f_{IB} \cdot f_{satB} \cdot \exp\left[\frac{V_{bei}}{n_B V_T}\left(1 - \frac{\delta_B \Theta P_d}{n_B}\right) - \frac{I_B R_{bb}}{n_B V_T}\right] \tag{3.5}$$

where $f_{IB} = \exp(a_{IB} I_C)$, and $f_{satB} = \exp(a_{satB} V_{bcirev})$, where a_{IB}, a_{SHB}, and a_{satB} are empirical parameters for the base current source, analogous to the collector current source parameters, n_B is the ideality factor, and R_{bb} is an empirical resistance similar to R_{bc}. Note that the various parameters of the model are empirical computational parameters, and not the actual physical ones.

3.2.2 Current Source Model Parameter Extraction

The empirical current source models described above have been formulated to enable a simple, fast, and direct parameter extraction strategy. Equation 3.4, which describes the collector current source, can be cast in linear form to yield Equation 3.6, using n number of sampled data at various biases, at a particular ambient temperature. Since the number of unknown parameters (matrix X) is 9, for $n > 9$, the solution of Equation 3.6 using $X = (A^T A)^{-1} A^T B$ minimizes the average squared-error between measured and modeled data. Thus, this procedure eliminates the post-extraction optimization step, which is generally found to be necessary to obtain an accurate correspondence between measured and modeled data in traditional extraction methods. Similarly, Equation 3.5 is cast in linear form (Equation 3.7), to obtain the parameters of the I_{BE} diode. The parameters of the current source model are extracted at different ambient temperatures and fitted into polynomial functions of ambient temperature, to predict the characteristics at all temperatures in the range of interest.

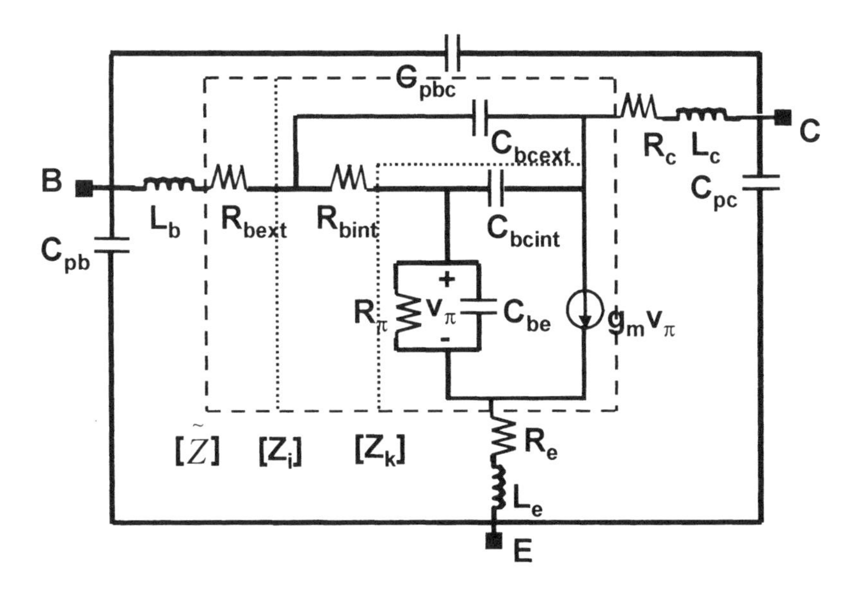

Figure 3.2. Equivalent small-signal model.

$$AX = B \tag{3.6}$$

$$A = \begin{bmatrix} 1 & P_{d,1} & P_{d,1}^2 & I_{C,1} & V_{cei,1} & V_{bcirev,1} & \dfrac{V_{bei}}{V_{T\,1}} & V_{bei}\cdot P_{d,1} & \dfrac{I_B}{V_{T1}} \\ \cdot & \cdot & \cdot & \cdot & \cdot & \cdot & \cdot & \cdot & \cdot \\ \cdot & \cdot & \cdot & \cdot & \cdot & \cdot & \cdot & \cdot & \cdot \\ 1 & P_{d,n} & P_{d,n}^2 & I_{C,n} & V_{cei,n} & V_{bcirev,n} & \dfrac{V_{bei}}{V_{T\,n}} & V_{bei}\cdot P_{d,n} & \dfrac{I_B}{V_{Tn}} \end{bmatrix}_{(n\times 9)}$$

$$X = \begin{bmatrix} \ln(I_{SCO}) \\ a_{SH}\Theta_0 \\ a_{SH}\Theta_0 b \\ a_I \\ a_V \\ a_{sat} \\ 1/n_C \\ -\delta\Theta_0/n_C^2 V_T \\ -R_{bc}/n_C \end{bmatrix}_{(9\times 1)} \qquad B = \begin{bmatrix} \ln(I_C)_1 \\ \cdot \\ \cdot \\ \ln(I_C)_n \end{bmatrix}_{(n\times 1)}$$

$$CY = D \tag{3.7}$$

$$C = \begin{bmatrix} 1 & P_d + bP_{d1}^2 & I_{C,1} & V_{bcirev,1} & V_{bei,1} & V_{bei} \cdot P_{d,1} & I_{B,1} \\ \bullet & \bullet & \bullet & \bullet & \bullet & \bullet & \bullet \\ \bullet & \bullet & \bullet & \bullet & \bullet & \bullet & \bullet \\ 1 & P_d + bP_{dn}^2 & I_{C,n} & V_{bcirev,n} & V_{bei,n} & V_{bei} \cdot P_{d,n} & I_{B,n} \end{bmatrix}_{(n\times 7)}$$

$$Y = \begin{bmatrix} \ln(I_{SBO}) \\ a_{SHB}\Theta_0 \\ a_{IB} \\ a_{satB} \\ 1/n_B V_T \\ -\delta_B \Theta_0 / n_B^2 V_T \\ -R_{bb}/n_B \end{bmatrix}_{(7\times 1)} \qquad D = \begin{bmatrix} \ln(I_B)_1 \\ \bullet \\ \bullet \\ \ln(I_B)_n \end{bmatrix}_{(n\times 1)}$$

3.2.3 Extraction of Intrinsic Capacitances

The base–collector and base–emitter capacitances are extracted from S-parameters measured at different bias points. To calculate the intrinsic and extrinsic base–collector capacitances (C_{bcint} and C_{bcext}, respectively), the expressions in Equations 3.8 and 3.9 are used. Here, the terms z_{ixy} are the Z-parameters of the block $[Z_{\mathrm{i}}]$ (see Figure 3.2), obtained after deembedding the extrinsic parasitic elements. Since the parameters of the hybrid π model show frequency dependence [17], in order to extract a lumped value for a given capacitance (at a given bias condition), Equations 3.8 and 3.9 are cast in linear form (Equations 3.10 and 3.11, respectively) using n number of measured data at different frequencies. Knowing the internal base resistance R_{bint}, we can obtain C_{bcext} and C_{bcint} from the solutions to Equations 3.10 and 3.11. The extraction procedure inherently minimizes the average squared error between the measured and modeled data, and thereby yields lumped values for the capacitances that minimize the average error between measured and modeled characteristics over the entire frequency range of interest. Similarly, C_{be} is extracted using Equation 3.13, which is the linear form of Equation 3.12. Here, the terms z_{kxy} are the Z-parameters of the block $[Z_{\mathrm{k}}]$ (see Figure 3.2), obtained after deembedding C_{bcext} and R_{bint} from $[Z_{\mathrm{i}}]$.

$$Re\left(\frac{1}{z_{i11}-z_{i12}}\right)=\frac{1}{\omega R_{bint}C_{bcint}}.Im\left(\frac{1}{z_{i22}-z_{i21}}\right) \tag{3.8}$$

$$Im\left(\frac{1}{z_{i22}-z_{i21}}\right)=\omega\left(C_{bcext}+C_{bcint}\right) \tag{3.9}$$

$$\begin{bmatrix}\omega_1\\ \cdot\\ \cdot\\ \omega_n\end{bmatrix}\left[\left(C_{bcext}+C_{bcint}\right)\right]=\begin{bmatrix}Im\left(\frac{1}{z_{i22}-z_{i21}}\right)_1\\ \cdot\\ \cdot\\ Im\left(\frac{1}{z_{i22}-z_{i21}}\right)_n\end{bmatrix} \tag{3.10}$$

$$\begin{bmatrix}Re\left(\frac{1}{z_{i11}-z_{i12}}\right)_1\\ \cdot\\ \cdot\\ Re\left(\frac{1}{z_{i11}-z_{i12}}\right)_n\end{bmatrix}\left[R_{bint}C_{bcint}\right]=\begin{bmatrix}\frac{1}{\omega_1}Im\left(\frac{1}{z_{i22}-z_{i21}}\right)_1\\ \cdot\\ \cdot\\ \frac{1}{\omega_n}Im\left(\frac{1}{z_{i22}-z_{i21}}\right)_n\end{bmatrix} \tag{3.11}$$

$$\omega.C_{be}=\mathrm{Im}\left(j\omega C_{bc\,\mathrm{int}}\frac{z_{k22}-z_{k12}}{z_{k12}}\right) \tag{3.12}$$

$$\begin{bmatrix}\omega_1\\ \cdot\\ \cdot\\ \omega_n\end{bmatrix}\left[C_{be}\right]=\begin{bmatrix}\mathrm{Im}\left(j\omega_1 C_{bc\,\mathrm{int}}\left(\frac{z_{k22}-z_{k12}}{z_{k12}}\right)_1\right)\\ \cdot\\ \cdot\\ \mathrm{Im}\left(j\omega_n C_{bc\,\mathrm{int}}\left(\frac{z_{k22}-z_{k12}}{z_{k12}}\right)_n\right)\end{bmatrix} \tag{3.13}$$

3.2.4 Extraction of Base Resistance

Accurate extraction of the base resistance is very important for accurate internal current source modeling. In a device where the base electrode is not distributed, the base resistance can be extracted using the traditional analytic expression: $\mathrm{Re}\{z_{11} - z_{12}\}$. However, in a device where the base resistance is distributed into external (R_{bext}) and internal (R_{bint}) component, this expression does not work very well. For such a case, the following expression is used for the extraction of the total base resistance [22]:

$$Z_Q \equiv \tilde{z}_{11} - \tilde{z}_{12}\frac{\tilde{z}_{21} - \tilde{z}_{12}}{\tilde{z}_{22} - \tilde{z}_{12}} \tag{3.14}$$

$\tilde{z}_{\mathrm{ij}}$ are the Z-parameters after deembedding the pad parasitics, except the external base resistance (R_{bext}), as shown in Figure 3.2. Equation 3.14 can be simplified as shown in Equation 3.15. The final expression in Equation 3.15 can be represented as an equivalent circuit, as shown in Figure 3.3. At high frequency, Z_π ($\equiv Z_{\mathrm{p}}$) goes to $\{1/(j\omega C_\pi)\}$ ($C_\pi \equiv C_{\mathrm{p}}$). With the assumption that $R_{\mathrm{bint}} \ll |X_{\mathrm{bc\_tot}}|$ over the frequency range of interest, we get the expression for R_{btotal} given in Equation 3.16.

$$\begin{aligned}
Z_Q &\equiv \tilde{z}_{11} - \tilde{z}_{12}\frac{\tilde{z}_{21} - \tilde{z}_{12}}{\tilde{z}_{22} - \tilde{z}_{12}} \\
&= R_{bext} + \frac{R_{bint}X_{bcext}}{R_{bint} + X_{bcext} + X_{bcint}} + \frac{R_{bint}X_{bcint}}{(1 + g_m Z_p)(R_{bint} + X_{bcext} + X_{bcint})} + \frac{Z_p}{1 + g_m Z_p} \\
&\quad - \left[\frac{R_{bint}X_{bcint}}{(1 + g_m Z_p)(R_{bint} + X_{bcext} + X_{bcint})} + \frac{Z_p}{1 + g_m Z_p}\right](-g_m Z_p) \\
&= R_{bext} + \frac{R_{bint}X_{bcext} + R_{bint}X_{bcint}}{R_{bint} + X_{bcext} + X_{bcint}} + \frac{R_p X_p}{R_p + X_p} \\
&= R_{bext} + R_{bint} \,//\, X_{bc_tot} + R_\pi \,//\, X_\pi \\
&= R_{bext} + R_{bint} \,//\, X_{bc_tot} + Z_\pi
\end{aligned} \tag{3.15}$$

$$R_{btotal} \approx real\left\{\tilde{z}_{11} - \tilde{z}_{12}\frac{\tilde{z}_{21} - \tilde{z}_{12}}{\tilde{z}_{22} - \tilde{z}_{12}}\right\}_{at\ high\ freq.} \tag{3.16}$$

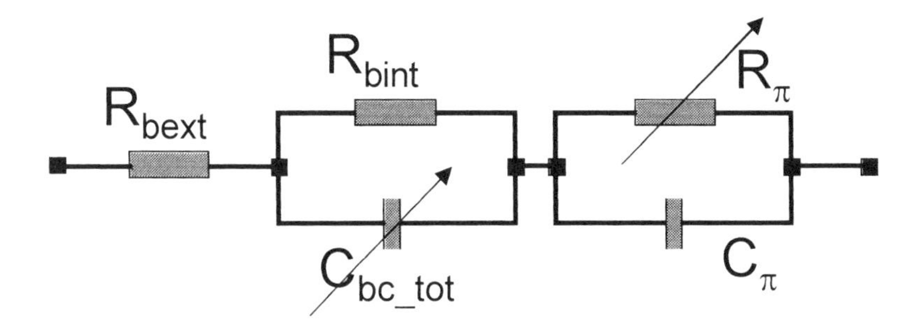

Figure 3.3. Equivalent-circuit representation of Equation 3.15.

The assumption $R_{\mathrm{bint}} << |X_{\mathrm{bc\_tot}}|$ can be justified beginning with the well-known expression for $f_{\max}$:

$$f_{\max} = \sqrt{\frac{f_t}{8\pi\ R_{btot} C_{bc\_tot}}} \tag{3.17}$$

Taking R_{bint} to be a fraction α of the total base resistance R_{btot} we get

$$\frac{|X_{bctot}|}{R_{bint}} = 4\frac{f_{\max}^2}{\alpha\ f f_t} \tag{3.18}$$

In the worst case, that is, at the cut-off frequency, and with $\alpha = 1$, $\mathrm{Re}\{R_{bint} \| X_{\mathrm{bctot}}\} = 0.94 R_{\mathrm{bint}} \cong R_{\mathrm{bint}}$. If $\alpha = 0.5$, this approximation gets better: $\mathrm{Re}\{R_{bint} \| X_{\mathrm{bctot}}\} = 0.985 R_{\mathrm{bint}}$.

To illustrate the robustness of this expression compared to the traditional $Z_{\mathrm{A}} = z_{11}\text{-}z_{12}$, typical values for the parameters were inserted in a bipolar small-signal model, the total base resistance was arbitrarily partitioned into external and internal parts in different ratios, and the expressions Z_{Q} and Z_{A}, evaluated. The results are shown in Figure 3.4. While Z_{Q} yields the same value for R_{btot} (= 9.8 Ω) each time, which is close to the value of 9.4 Ω that we started with, $\mathrm{Re}\{z_{11}\text{-}z_{12}\}$ yields different values for different ratios of external and internal base resistance.

3.2.5 Parameter Extraction Procedure

The model parameter extraction procedure begins with the extraction of the access parasitic terms: the pad capacitances, and the access inductances and resistances. This can done from the measured *S*-parameters of either deemebedding short and open test structures, or directly from the device-under-test (DUT) structure, using the open-collector–cold S-parameter measurement method [23,24]. The extracted parasitic elements are deembedded from the measured S-parameters to yield the instrinsic S (or Y or Z) parameters. The total base resistance is extracted, after deembedding, from the $[\tilde{z}]$ matrix (see Figure 3.2) using the base resistance extraction technique described in the previous section. The internal node voltages obtained after deembedding the resistive parasitics are used to model the internal current sources using the technique described in Section 3.2.2. The base–collector and base–emitter capacitances are extracted using the procedure described in Section 3.2.3, to complete the model parameter extraction.

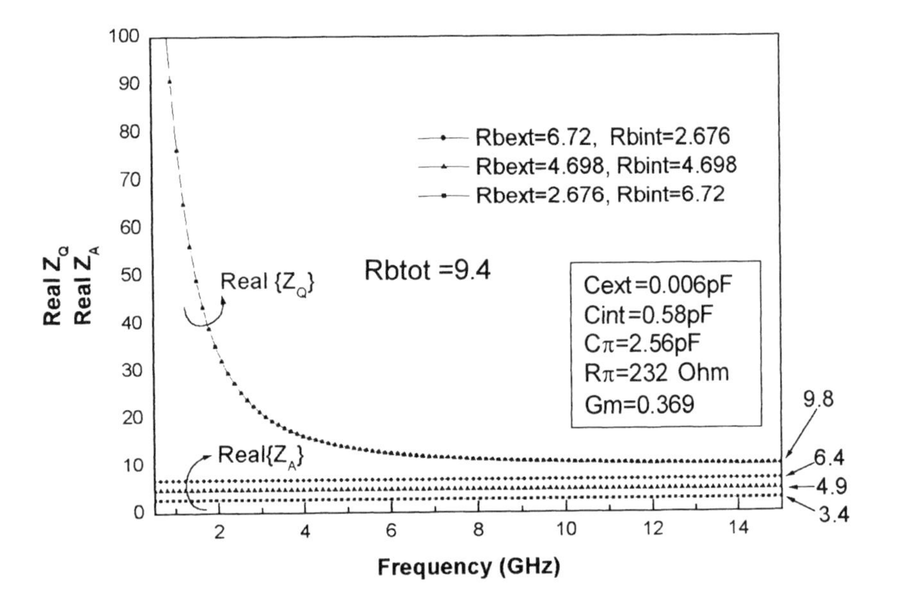

Figure 3.4. Comparsion of base resistance extraction using the method of expression Z_Q and the traditional method of expression Z_A.

3.3 TEMPERATURED-DEPENDENT INGAP/GAAS HBT LARGE-SIGNAL MODEL

This section illustrates the extraction and verification of the large-signal model described in the previous section for the case of the InGaP/GaAs HBT. The particular device being modeled is a two-finger, 2 × 40-μm emitter-area (per finger) InGaP/GaAs HBT. First, the parasitic capacitances (C_{pb}, C_{pc}, C_{pbc}), the emitter elements (R_e, L_e), extrinsic base elements (R_{bext}, L_b) and collector parasitic components (R_c, L_c) are extracted using the open-collector–cold S-parameter measurement method, and deembedded.

The total base resistance is extracted after deembedding, from the $[\tilde{z}]$ matrix (see Figure 3.2) using the technique (Equation 3.16) described in Section 3.2.4. Figure 3.5 shows the frequency behavior of Equation 3.16 at several bias points. It is observed from the figure that all the curves converge to approximately the same value (6 Ω), which is the extracted value of base resistance.

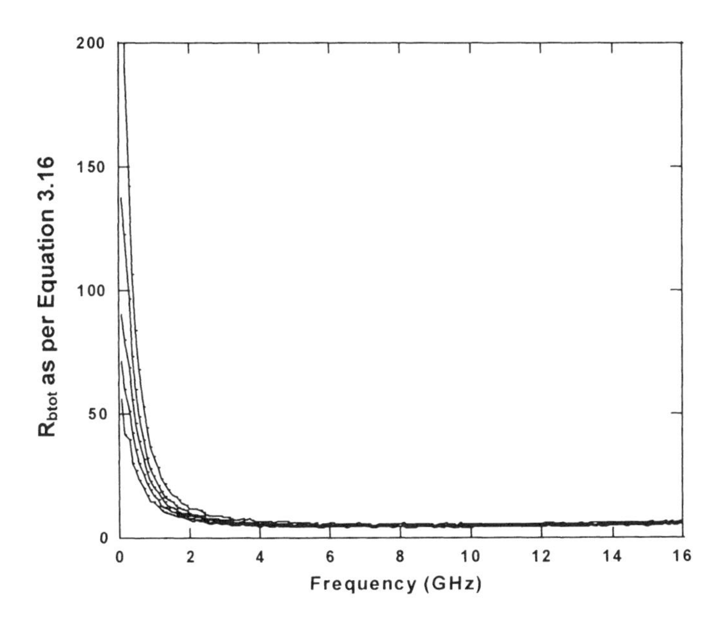

Figure 3.5. Extraction of R_{btot} from Equation 3.16, with V_{CE} = 2.5 V and I_B = 100, 150, 200, 250, and 300 μA.

The resistive parasitics extracted above are deembedded from the measured I–V data to obtain the internal node voltages, and the parameters of the current sources are extracted using Equations 3.6 and 3.7. The current source model parameters are extracted at different ambient temperatures between 25°C and 100°C, and fitted into polynomial functions of ambient temperature. The measured versus modeled I–V, and V_{be}–V_{ce} characteristics (at 25°C and 100°C) are shown in Figures 3.6 and 3.7, respectively. The extracted parameters at different ambient temperatures T_{amb} are summarized in Tables 3.1 and 3.2. To complete the verification of the current source model, the values of the model parameters are computed at a temperature at which they were not extracted (in this case, 85°C), and the resulting modeled results are verified against measured data (Figures 3.8 and 3.9). The mean (averaged over the V_{ce} range, 0–6 V) absolute percentage error between measured and modeled DC collector current I_C at different base bias currents and temperatures is shown in Figure 3.10, and indicates excellent accuracy (maximum error of less than 1%) across this temperature range.

From the current source model, the corresponding small-signal parameters, such as, base–emitter dynamic resistance and transconductance, are directly obtained under small-signal operation. The remaining elements, namely the base-collector and base–emitter capacitances, are extracted from S-parameters measured at different bias points, using Equations 3.10 and 3.11, and 3.13, respectively. The extracted parasitic elements and intrinsic equivalent circuit elements at a subset of biases are listed in Table 3.3. Table 3.3 also lists the residual error between measured and modeled S-parameters (over frequency f), at the respective bias conditions (which quantifies the accuracy of the small-signal model), calculated according to the following equation:

$$Error = \sum_{i,j=1}^{2} \sum_{k=1}^{N} \frac{\left| S_{ij}^{\text{mod}}(f_k) - S_{ij}^{meas}(f_k) \right|}{\max_k \left| S_{ij}^{meas}(f_k) \right|} \tag{3.19}$$

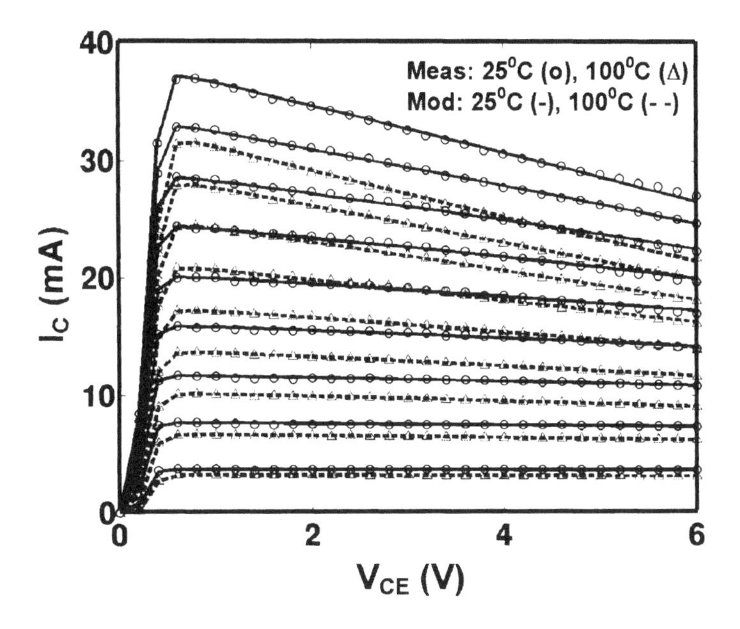

Figure 3.6. Measured [25°C (o), 100°C (Δ)] versus modeled [25°C (-), 100°C (--)] DC I–V characteristics; I_B = 50–450 μA in 50-μA increments.

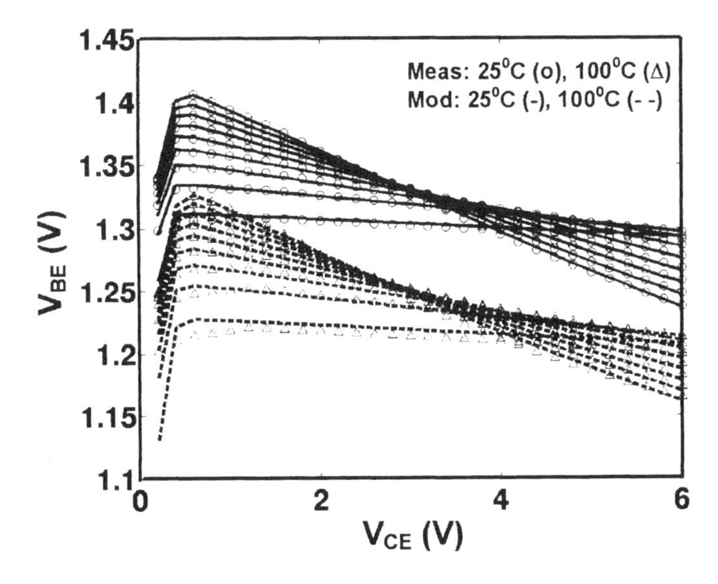

Figure 3.7. Measured [25°C (o), 100°C (Δ)] versus modeled [25°C (-), 100°C (--)] V_{BE}–V_{CE} characteristics; I_B = 50–450 μA in 50-μA increments.

Table 3.1. Extracted parameters for the I_{CC} current source

T_{amb} (°C)	I_{SC0} (A)	$a_{SH}\Theta_0$ (W^{-1})	b (W^{-1})	a_I (A^{-1})	a_V (V^{-1})	a_{sat} (V^{-1})	n_C	$\delta\Theta_0$ (W^{-1})	R_{bc} (Ω)
25	5.0e-23	-204.2	-0.393	10.11	-0.0122	0.041	1.101	4.08	0.03
50	2.3e-21	-199.3	-0.436	10.62	-0.0127	0.039	1.180	4.62	0.30
75	6.7e-20	-191.8	-0.473	10.65	-0.0127	0.036	1.256	5.09	1.09
100	1.3e-18	-180.9	-0.495	10.58	-0.0110	0.032	1.332	5.44	1.80

Table 3.2. Extracted parameters for the I_{BE} current source

T_{amb} (°C)	I_{SB0} (A)	$a_{SHB}\Theta_0$ (W^{-1})	a_{IB} (A^{-1})	a_{satB} (V^{-1})	n_B	$\delta_B\Theta_0$ (W^{-1})	R_{bb} (Ω)
25	4.5e-22	-122.0	-39.8	0.186	1.283	3.0	136.3
50	7.3e-20	-94.3	-41.5	0.225	1.446	2.74	155.8
75	1.0e-17	-61.5	-42.8	0.268	1.652	1.87	178.2
100	7.3e-16	-27.8	-44.5	0.309	1.889	4.13	201.9

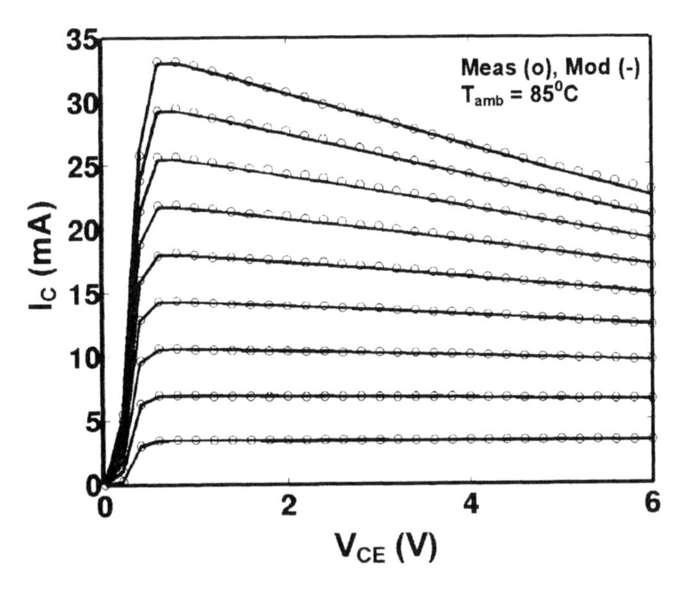

Figure 3.8. Measured (o) versus modeled (-) DC I–V characteristics at an ambient temperature of 85°C; I_B = 50–450 μA in 50-μA increments.

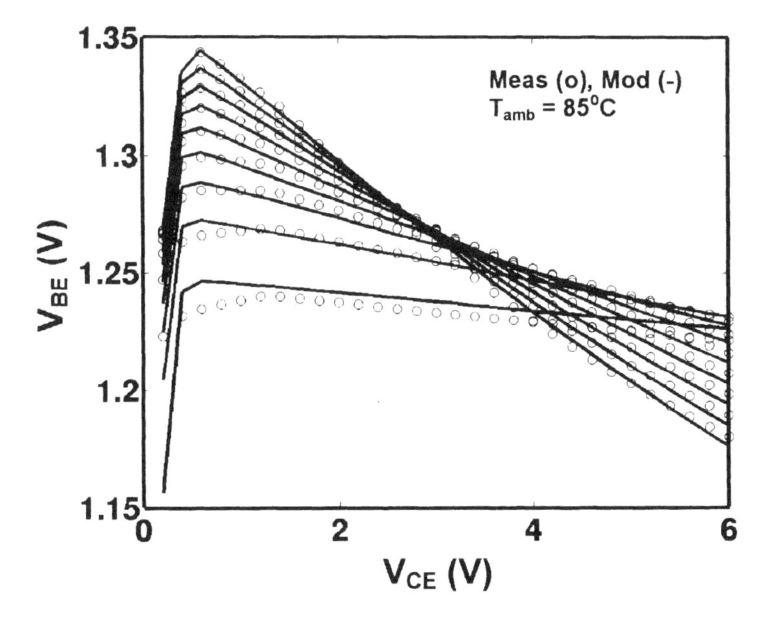

Figure 3.9. Measured (o) versus modeled (-) V_{BE}–V_{CE} characteristics at an ambient temperature of 85°C; I_B = 50–450 μA in 50-μA increments.

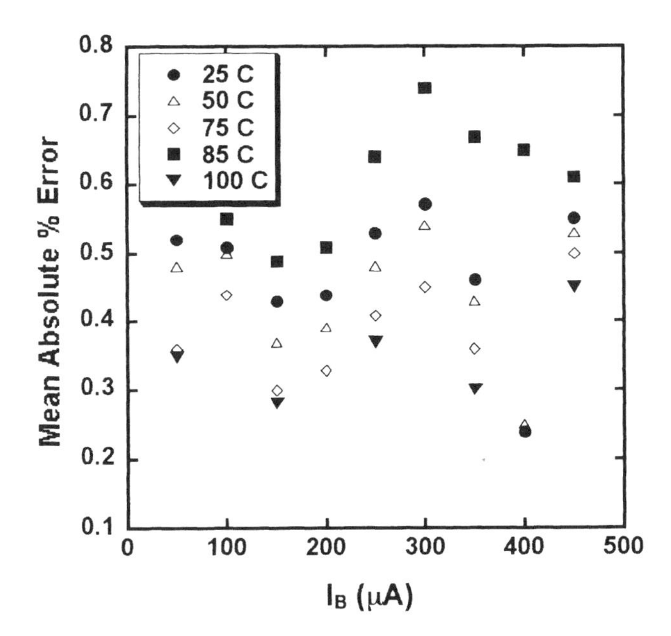

Figure 3.10. Mean absolute percentage error (averaged over V_{CE} =0–6 V) versus I_B at different ambient temperatures.

The extracted intrinsic capacitances are fitted into polynomial functions of internal voltages and currents. Temperature coefficients of the intrinsic elements are typically small (as has also been observed in other studies [5,7,25]), and a linear ambient-temperature dependence or a value averaged over the ambient temperature range is found to suffice. Figures 3.11 and 3.12 show the correspondence between the measured and the modeled S-parameters at different bias conditions and ambient temperatures.

Table 3.3. Extracted parasitic and intrinsic parameters (V_{CE} = 3 V)

C_{pb} (fF)	C_{pc} (fF)	C_{pbc} (fF)	R_{bext} (Ω)	R_c (Ω)	R_e (Ω)	L_b (pH)	L_c (pH)	L_e (pH)
50	72	4	1.5	1.1	1.4	123	60	37

I_C (mA)	R_{btot} (Ω)	R_{bint} (Ω)	C_{bcext} (fF)	C_{bcint} (fF)	C_{be} (pF)	S-Parameter Error (%)
5	6.0	4.5	47.6	25.3	0.31	2.7
10	5.9	4.4	50.4	23.1	0.35	3.3
15	6.0	4.5	41.5	28.2	0.43	2.9
20	5.8	4.3	43.3	35.4	0.53	3.8

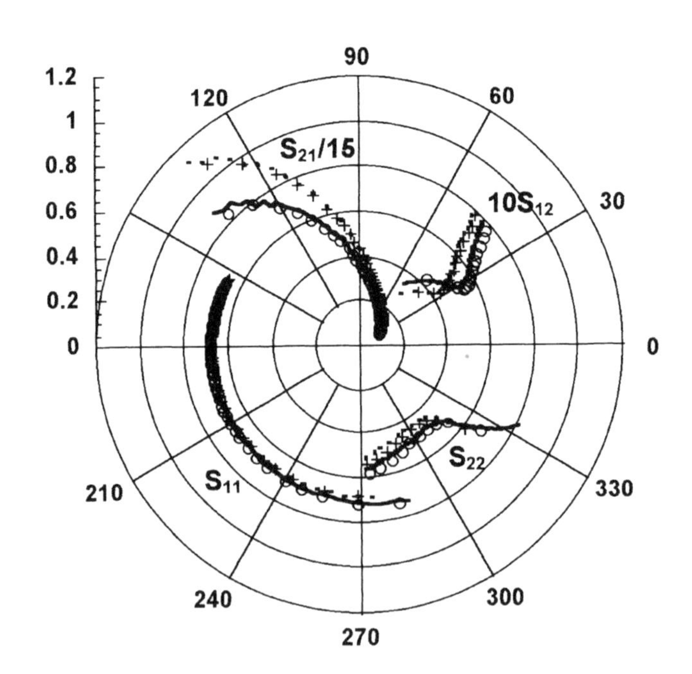

Figure 3.11. Measured [I_C=10 mA (o), I_C=15 mA (+)] versus modeled [I_C=10 mA (-), I_C=15 mA (--)] S-parameters (1–20 GHz) at V_{CE}=3 V, at 25°C.

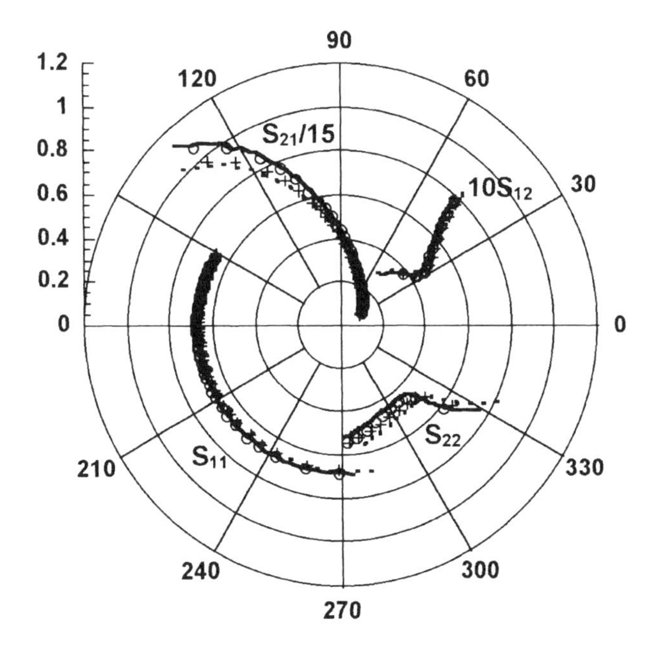

Figure 3.12. Measured [T_{amb}=25°C (o), T_{amb}=100°C (+)] versus modeled [T_{amb}=25°C (-), T_{amb}=100°C (--)] S-parameters (1–20 GHz) at V_{CE}=3 V and I_C=15 mA.

To verify the large-signal performance of the model, load- and source-pull measurements were performed, and compared with the modeled performance using harmonic balance simulations. The large-signal model was tested up to the C band, and was found to show good agreement with measured large-signal fundamental and harmonic power characteristics. Figure 3.13 demonstrates the correlation between measured and modeled output power, gain, and power-added efficiency at a selection of biases and ambient temperatures. To verify the harmonic prediction capability of the model, power-sweep measurements were performed with 50-Ω terminations. Figure 3.14 demonstrates the correspondence between measured and modeled harmonic power characteristics. Thus, it is seen that this model is able to accurately predict the DC, small-signal, and large-signal performance of the device over a wide range of biases and temperatures.

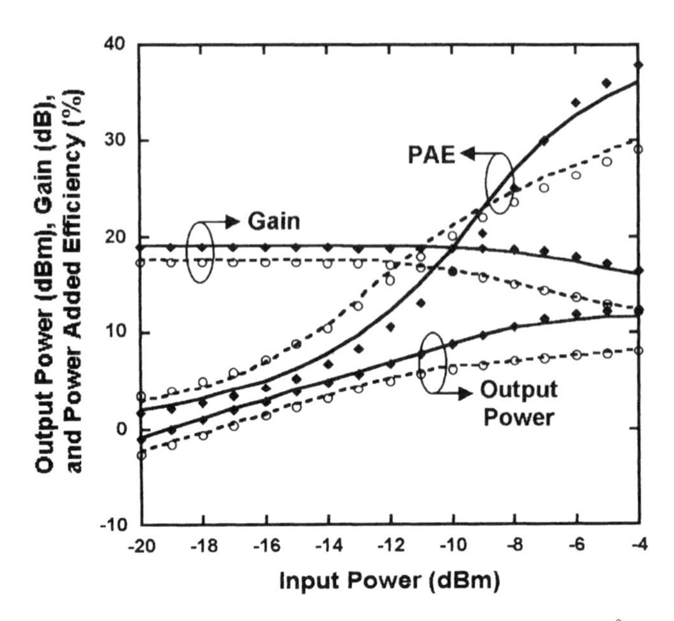

Figure 3.13. Measured [I_C=12 mA, T_{amb}=25°C (o); I_C=6 mA, T_{amb}=85°C (◆)] versus modeled [I_C=12 mA, T_{amb}=25°C (-); I_C=6 mA, T_{amb}=85°C (--)] output power, gain, and power-added efficiency, versus input power at V_{CE} = 3 V, at 5.8 GHz. $\Gamma_S = 0.57 \angle 169.5^{\circ}$, and $\Gamma_L = 0.63 \angle 38.9^{\circ}$ for the I_C=12m A, T_{amb}=25°C bias condition; and $\Gamma_S = 0.64 \angle 147.4^{0}$, and $\Gamma_L = 0.63 \angle 40.6^{0}$ for the I_C=6 mA, T_{amb}=85°C bias condition.

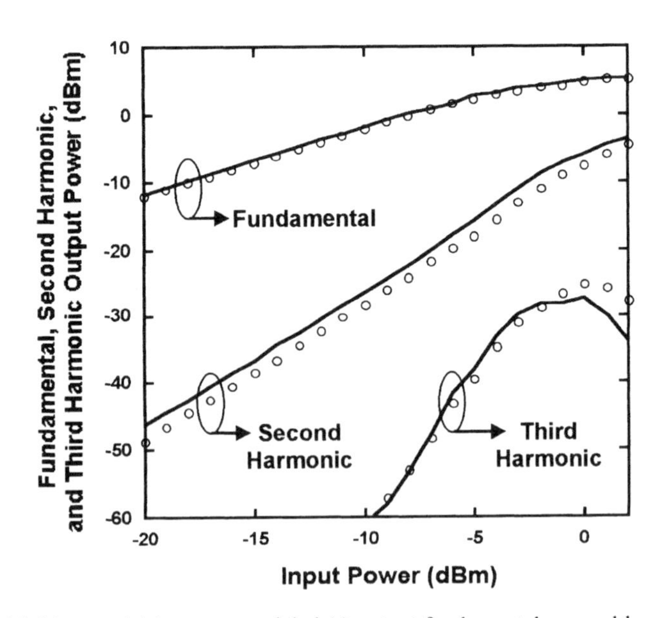

Figure 3.14. Measured (o) versus modeled (-) output fundamental, second-harmonic and third-harmonic power versus input power (with 50 Ω source and load terminations) at V_{CE} = 3 V and I_C = 12 mA, at 5.8 GHz, at 25°C.

3.4 EMPIRICAL SI BJT LARGE-SIGNAL MODEL

The large and small signal model topologies for the Si BJT model are similar to those of the GaAs HBT. The large-signal equivalent-circuit model of the Si BJT is shown in Figure 3.15. In the case of the Si BJT, a simplified current source model formulation suffices. The Early voltage can be extracted directly from the slope of the DC I–V curves. Also, a simplified thermal resistance model, with a thermal resistance that is constant over ambient temperature, is used. The ideality factor is assumed to be constant, and the temperature dependence of the characteristics is accounted for by the thermal resistance and an ambient temperature-dependent term in the preexponential factor. The current source models used are shown below

$$I_{CC} = I_{SCC}\left[\exp\left(\frac{V_{bei}}{NV_{TCC}}\right)-1\right]\left(1+\frac{V_{cei}}{V_A}\right) \tag{3.20}$$

$$I_{EE} = A_r I_{SBC}\left[\exp\left(\frac{V_{bci}}{NV_{TBC}}\right)-1\right] \tag{3.21}$$

$$I_{BE} = I_{SBE}\left[\exp\left(\frac{V_{bei}}{NV_{TBE}}\right)-1\right] \tag{3.22}$$

$$I_{BC} = I_{SBC}\left[\exp\left(\frac{V_{bci}}{NV_{TBC}}\right)-1\right] \tag{3.23}$$

where V_A is the Early voltage, V_{xyi} are internal voltages, and I_{SXY} are the temperature dependent parameters given by

$$I_{SXY} = I_{SXY0}\exp[A_{XY}(R_{th}P_d + T_A - T_{A0})] \tag{3.24}$$

where P_d is the power dissipation, T_A is the actual ambient temperature, T_{A0} is the reference ambient temperature, R_{th} is the thermal resistance, and A_{XY} are parameters for the temperature dependence of the terms I_{SXY}. The current sources I_{CC} and I_{BE} are dominant in the normal active region of the DC I–V curves above the knee voltage. In this region, I_{CC} and I_{BE} represent most of collector and base terminal currents, I_{ct} and I_{bt}, respectively ($I_{ct} \approx I_{CC}$) and $I_{bt} \approx I_{BE}$). As before, Equation 3.20 can be converted to a linear equation, using n number of sampled data:

$$AX = B \tag{3.25}$$

$$A \equiv \begin{bmatrix} 1 & V_{bei,1} & P_{diss,1} & \left(T_{A,1} - T_{A0}\right) \\ 1 & V_{bei,2} & P_{diss,2} & \left(T_{A,2} - T_{A0}\right) \\ \vdots & \vdots & \vdots & \vdots \\ 1 & V_{bei,n} & P_{diss,n} & \left(T_{A,n} - T_{A0}\right) \end{bmatrix}_{n\times 4},$$

$$X \equiv \begin{bmatrix} \ln\left(I_{SCC}\right) \\ NV_{TCC}^{-1} \\ R_{th}A_{CC} \\ A_{CC} \end{bmatrix}_{4\times 1}, \quad B \equiv \begin{bmatrix} \ln\left[\dfrac{I_{ct,2}}{1+V_A^{-1}V_{cei,2}}\right] \\ \ln\left[\dfrac{I_{ct,2}}{1+V_A^{-1}V_{cei,2}}\right] \\ \ln\left[\dfrac{I_{ct,n}}{1+V_A^{-1}V_{cei,n}}\right] \end{bmatrix}_{n\times 1}$$

Equation 3.25 can be constructed from the measured DC I–V data in the normal active region at several ambient temperatures. For $n>4$, it is an overdetermined linear equation, solving which the parameters NV_{TCC}, I_{SCC0}, A_{CC}, and R_{th} are extracted.

By a similar procedure, a linear equation (Equation 3.26) for the current source I_{BE} can be constructed, from which the parameters NV_{TBE}, I_{SBE0}, and A_{BC} are extracted:

$$CY = D \tag{3.26}$$

$$Y \equiv \begin{bmatrix} \ln\left(I_{SBE}\right) \\ NV_{TBE}^{-1} \\ A_{BE} \end{bmatrix}_{3\times 1}, \quad D \equiv \begin{bmatrix} \ln\left(I_{bt,1}\right) \\ \vdots \\ \ln\left(I_{bt,m}\right) \end{bmatrix}$$

$$C \equiv \begin{bmatrix} 1 & V_{bei,1} & R_{th}P_{d,1} + T_{A,1} - T_{A0} \\ \vdots & \vdots & \vdots \\ 1 & V_{bei,1} & R_{th}P_{d,1} + T_{A,1} - T_{A0} \end{bmatrix}_{m\times 3}$$

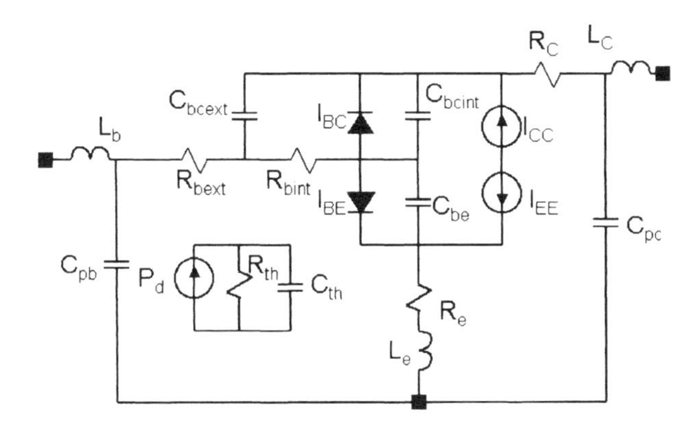

Figure 3.15. Large-signal equivalent circuit for the Si BJT.

The parameters of the current sources I_{EE} and I_{BC}, which are dominant in the reverse bias mode, are extracted from the reverse Gummel plot.

The model described here is extracted for a five-finger 0.4 × 20-μm-emitter Si BJT for practical verification. The base resistance is extracted using the method discussed in Section 3.2.4. Figure 3.16 shows the frequency behavior of Equation 3.16 at various bias points. The other access parasitic terms are extracted using the open collector and cold S-parameter method. The extracted extrinsic parameters, and the parameters for the current sources are listed in Table 3.4. The current source model parameters in Table 3.4 are determined without any optimization or trimming process, and are extracted directly from the measured data. Figures 3.17 and 3.18 show the measured and modeled collector currents and base–emitter terminal voltages at 0°C, 25°C, and 60°C. The intrinsic capacitances are extracted at different biases and fitted into polynomial functions of the internal voltages and currents. Figure 3.19 shows the measured and modeled S-parameters at the bias condition $I_B = 80$ μA, $V_{CE} = 2.5$ V, at an ambient temperature of 60°C.

To verify the large-signal performance of the model, load- and source-pull measurements were performed at several ambient temperatures. As shown in Figure 3.20, there is a good correlation between the measured and modeled power, gain, and efficiencies.

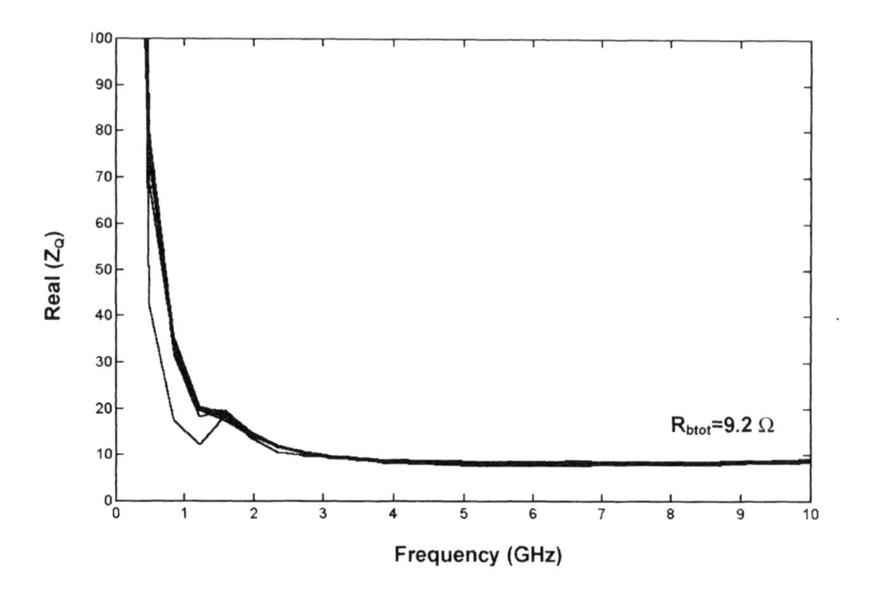

Figure 3.16. Extraction of R_{btot} using Equation 3.16; V_{CE} = 2.5 V, and I_B = 35, 60, 85, 110, 135, and 160 μA.

Table 3.4. Extracted parameters for the Si-BJT current source model.

C_{pb} (fF)	116	NV_{TBC} (V)	0.031
C_{pc} (fF)	158	A_{CC} (K^{-1})	0.042
R_e (Ω)	6.1	A_{BE} (K^{-1})	0.042
L_e (pH)	21	A_{BC} (K^{-1})	0.041
R_{btot} (Ω)	9.2	I_{SCC0} (A)	1.17×10^{-13}
R_c (Ω)	2.6	I_{SBE0} (A)	1.51×10^{-16}
T_{A0} (°C)	25	I_{SBC0} (A)	2.12×10^{-15}
V_A (V)	15.6	A_r	0.37
NV_{TCC} (V)	0.035	R_{th} (kW^{-1})	231.3
NV_{TBE} (V)	0.032		

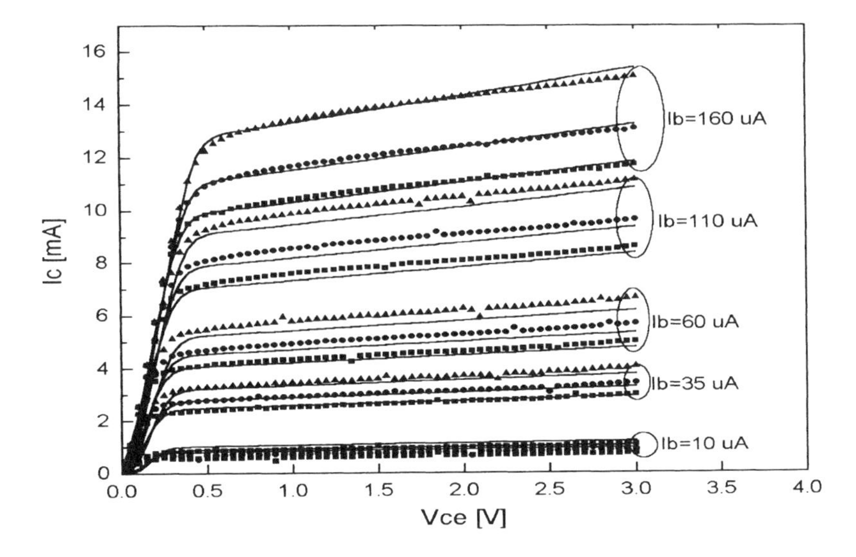

Figure 3.17. Modeled (-) versus measured (■: 0°C, ●: 25°C, and ▲: 60°C) DC *I–V* characteristics.

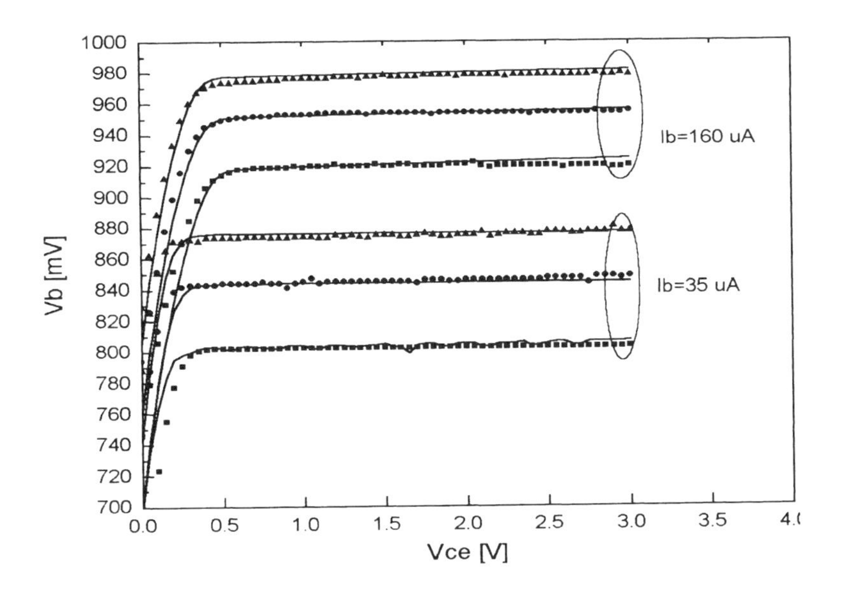

Figure 3.18. Modeled (-) versus measured (■: 0°C, ●: 25°C, and ▲: 60°C) V_B–V_{CE} characteristics.

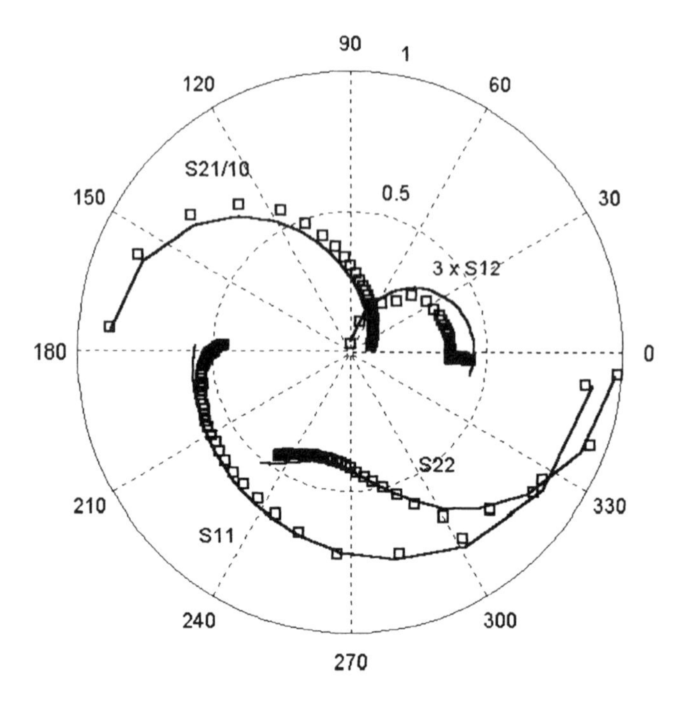

Figure 3.19. Modeled (-) versus measured (□) S-parameters (45 MHz – 15 GHz) at the bias condition $I_B = 80\ \mu A$, and $V_{CE} = 2.5$ V, at 60°C.

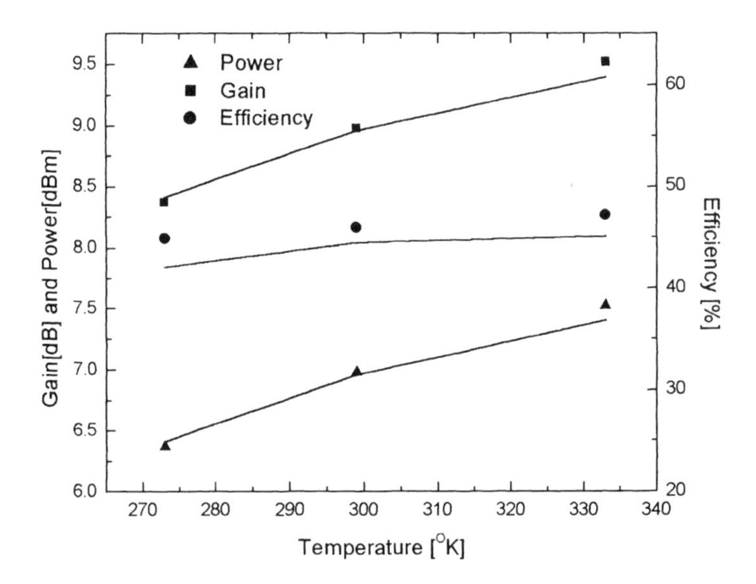

Figure 3.20. Modeled (-) and measured output power (▲), gain (■), and efficiency (●) at $I_B = 70\ \mu A$, and $V_{CE} = 2.5$ V, versus temperature at an input power level of –2 dBm at 2.4 GHz.

3.5 EXTENSION OF THE EMPIRICAL MODELING METHOD TO THE SIGE HBT

The empirical modeling method developed in this chapter has been demonstrated for GaAs HBT and Si BJT devices. In this section, the extension of the technique to the case of the SiGe HBT is explored. The major focus is on accurate temperature-dependent modeling of the nonlinear current sources of high-f_T SiGe HBTs, with emphasis on simple and fast parameter extraction techniques. SiGe HBTs are quite challenging to model accurately. Figure 3.21 shows the measured forward DC I–V characteristics of the SiGe HBT device under consideration, which has emitter dimensions 0.8 × 10 μm, at 25°C and 100°C. The strong temperature dependence of the DC I–V characteristics can clearly be seen from the data. In addition, the presence of the self-heating effect is also apparent in the negative differential resistance of the output characteristics. Because of the inherent tradeoff between f_T and breakdown voltage, BV_{CEO} is relatively low for this device (1.8 V in this case). Therefore, significant avalanche currents are generated in the base–collector region at voltages lower than the typical supply voltage used in circuit applications. This avalanche current is dependent on both the bias current and bias voltage levels (Figure 3.21), and must be properly accounted for. Accurate modeling must additionally consider the Early effect, the bias dependence of the current gain, Kirk effect, and heterojunction barrier, as well as the non-silicon-like temperature dependences [3].

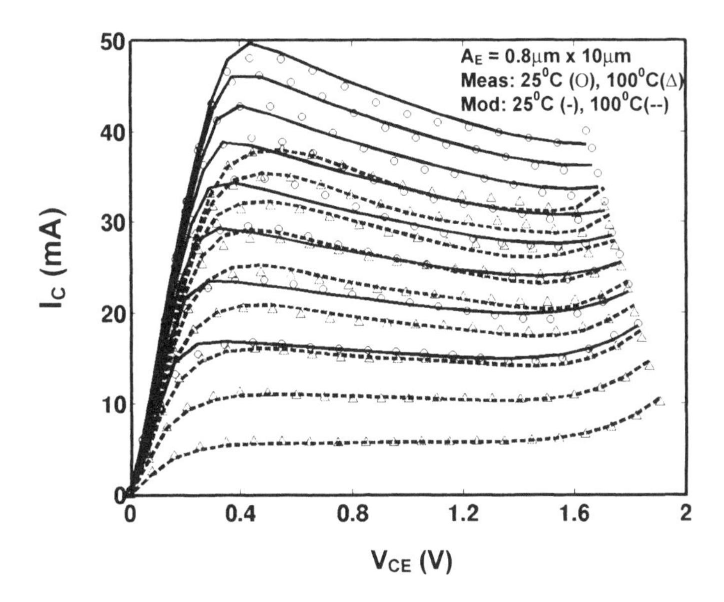

Figure 3.21. Measured [25°C (o), 100°C (Δ)] versus modeled [25°C (-), 100°C (--)] DC I–V characteristics; I_B=25–225 μA in 25-μA increments.

The empirical I_{CC} current source model is similar to that in the case of the GaAs HBT, with the addition of a few terms to account for effects that are not significant in the InGaP/GaAs HBT. The Early effect is one such effect. The Early effect is modeled by the factor $f_E = \exp(V_{cei}/V_A)$, where V_{cei} is the internal collector–emitter voltage, and f_E is an approximation of $(1+V_{cei}/V_A)$, assuming that (V_{cei}/V_A) is sufficiently small. The generated avalanche current I_{avl} is given by [26]:

$$I_{avl} = k\,(M-1)\,I_{Ci} \tag{3.27}$$

where I_{Ci} is the collector current without avalanche, and M is given by

$$M = \frac{1}{1-\left(\dfrac{V_{bci}}{V_{bco}}\right)^m} \tag{3.28}$$

where V_{cbi} is the internal base–collector voltage. The total collector current under weak avalanche is therefore $I_{Ci}+I_{avl}$, and this can be approximated by $I_{Ci}.\exp\{k(V_{cbi}/V_{cbo})^m\}$, assuming that V_{cbi} is significantly smaller than V_{cbo}. Now, using $a^x \approx (1+\ \ln x)$, the term above can further be approximated by $I_{Ci}.\exp\{k(1+m\ \ln V_{cb}/(V_{cbo})^m\}$. Absorbing the first term (i.e., $\exp\{k/(V_{cbo})^m\}$) into the preexponential factor I_{SC0} we get the avalanche factor $f_{avl} = \exp\{a_{avl}.\ln V_{bcibk}\}$, where $a_{avl} = m.k/(V_{cbo})^m$, and V_{bcibk} is defined as being equal to $-V_{bci}$ when $V_{bci} < 0$, and 1 otherwise. This definition of V_{bcibk} has been designed to overcome problems that would otherwise occur when taking the logarithm of a negative number. The dependence of the avalanche current on the bias current is modeled by the empirical factor $f_I = \exp(a_I I_C)$.

The empirical terms modeling the self-heating effect (f_{SH}), and the saturation region (f_{sat}) are identical to those used in the GaAs HBT model. The resulting model for the I_{CC} current source at a particular ambient temperature is given by

$$I_{CC} = I_{SC0}\cdot f_E\cdot f_{SH}\cdot f_{avl}\cdot f_I\cdot f_{sat}\cdot\exp\left(\frac{V_{bei}-I_B R_{bc}}{n_C V_T}\right) \tag{3.29}$$

All the terms in this equation vary with ambient temperature, including the preexponential factor I_{SCO}, since some effects, such as the avalanche current, are partially absorbed into this term. While the derivation of the various parameters in Equation 3.29 has been performed to account for particular physical effects in SiGe HBTs, other phenomena have also been empirically absorbed into, and are thus described by, these factors. For instance, f_E empirically facilitates the modeling of the Kirk effect and heterojunction barrier effect, which helps improve model accuracy at lower collector–emitter voltages and high bias current density. Similarly, f_I facilitates the modeling of the current gain at high bias current, analogous to the knee currents of the standard Gummel–Poon model, and contributes to the current dependence of the empirical resistance R_{bc}.

The model for the I_{BE} diode, from arguments similar to those in the case of the GaAs HBT, is given by

$$I_{BE} = I_{SBO} \cdot f_{SHB} \cdot f_{avlB} \cdot f_{IB} \cdot f_{satB} \cdot \exp\left(\frac{V_{bei} - I_B R_{bb}}{n_B V_T}\right) \tag{3.30}$$

where $f_{avlB} = \exp\{a_{avlB} \ln(V_{bcibk})\}$, $f_{IB} = \exp(a_{IB} I_B)$, and $f_{satB} = \exp(a_{satB} V_{bcirev})$, where a_{avlB}, a_{IB}, a_{SHB}, and a_{satB} are empirical parameters for the base current source, analogous to the collector current source parameters, n_B is the ideality factor; and R_{bb} is an empirical resistance similar to R_{bc}.

The extraction and verification of this SiGe HBT model was performed on a SiGe HBT device with emitter dimensions 0.8 × 10 μm. The large-signal equivalent circuit of the SiGe HBT is shown in Figure 3.22. The parasitic elements are extracted first, and appropriately deembedded. The parasitic capacitances (C_{pb} and C_{pc}), and the resistances, R_{pbe} and R_{pce}, which model the lossy nature of the silicon substrate, are extracted from an "open" test-structure. The remaining access parasitic terms (R_{bext}, R_c, R_e, L_b, L_c, and L_e) are directly extracted from the device test-structure, using open-collector measurements [23]. The internal base resistance is extracted from measured active S-parameters, using the technique described in Section 3.2.4.

The current source equations (Equations 3.29 and 3.30) are cast in linear form (Equations 3.31 and 3.32, respectively), the current source model parameters are extracted, similar to the GaAs HBT case, at different ambient temperatures between 25°C and 100°C, and fitted into polynomial functions of ambient temperature. The measured versus modeled I–V, and V_{be}–V_{ce} characteristics (at 25°C and 100°C) are shown in Figures 3.21 and 3.23, respectively. The extracted parameters at different ambient temperatures are summarized in Tables 3.5 and 3.6. To complete the verification of the current source model, the values of the model parameters are computed at a temperature at which they were not extracted (in this case, 85°C), and the resulting modeled results are verified against measured data (Figures 3.24 and 3.25). The average percentage error at different temperatures is shown in Figure 3.26, and indicates good accuracy across this temperature range.

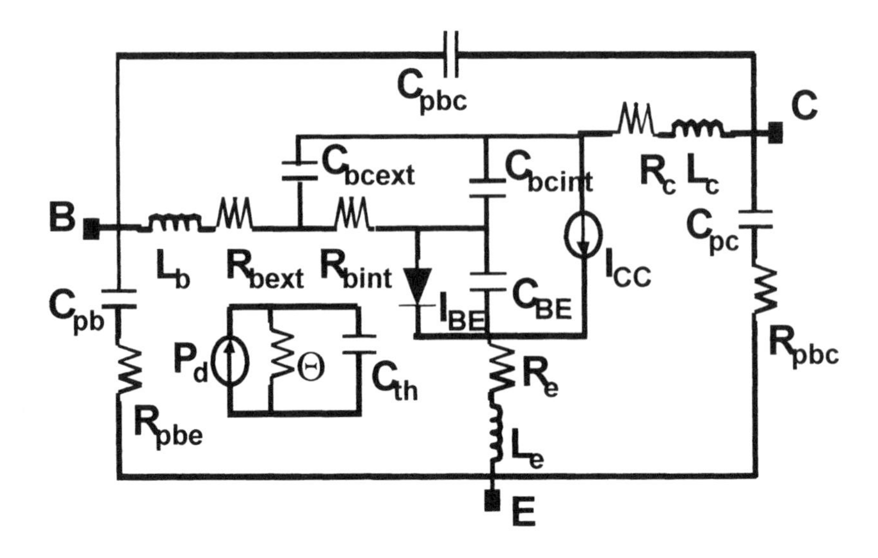

Figure 3.22. Equivalent circuit for the SiGe HBT CAD model.

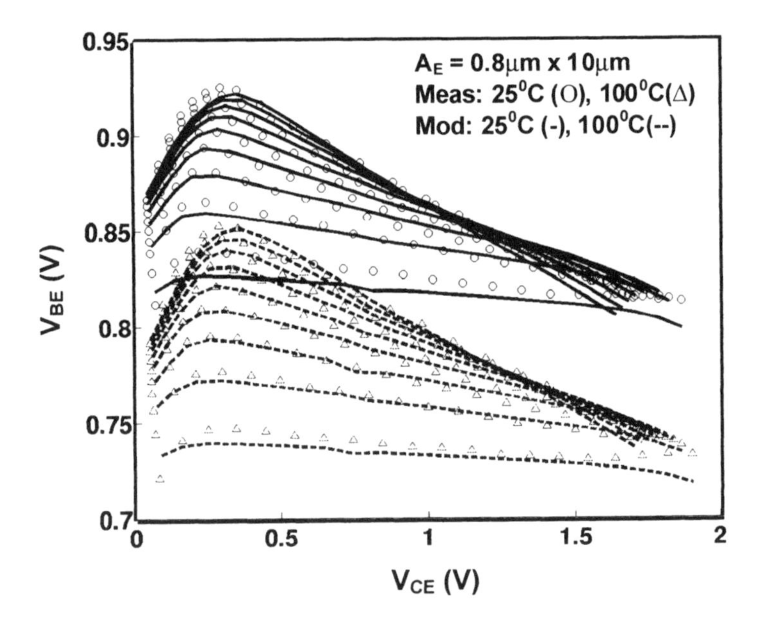

Figure 3.23. Measured [25°C (o), 100°C (Δ)] versus modeled [25°C (-), 100°C (--)] V_{BE}–V_{CE} characteristics; I_B=25–225 μA in 25-μA increments.

Table 3.5. Extracted parameters for current source I_{CC}.

T_{amb} (°C)	I_{SC0} (A)	V_A (V)	$a_{SH}\Theta_0$ (W^{-1})	b (W^{-1})	a_{avl}	a_l (A^{-1})	a_{sat} (V^{-1})	n_C	R_{bc} (Ω)
25	1.54e-14	6.16	37.06	7.84	0.0051	25.43	0.0075	1.182	137.44
50	3.56e-14	7.07	43.07	6.73	0.0060	24.88	0.0069	1.184	144.55
75	2.03e-13	6.60	43.27	7.19	0.0041	17.92	0.0067	1.228	152.50
100	1.35e-12	6.37	42.95	7.53	0.0069	16.82	0.0070	1.287	157.78

Table 3.6. Extracted parameters for current source I_{BE}.

T_{amb} (°C)	I_{SB0} (A)	$a_{SHB}\Theta_0$ (W^{-1})	a_{avlB}	a_{lB} (A^{-1})	a_{satB} (V^{-1})	n_B	R_{bb} (Ω)
25	6.50e-15	36.05	-0.0087	-14.40	0.07	1.428	190.3
50	7.39e-15	40.65	-0.0212	-17.83	0.10	1.389	166.5
75	1.19e-14	44.17	-0.0272	-29.66	0.09	1.366	151.4
100	3.20e-14	46.38	-0.0136	-36.92	0.04	1.374	148.3

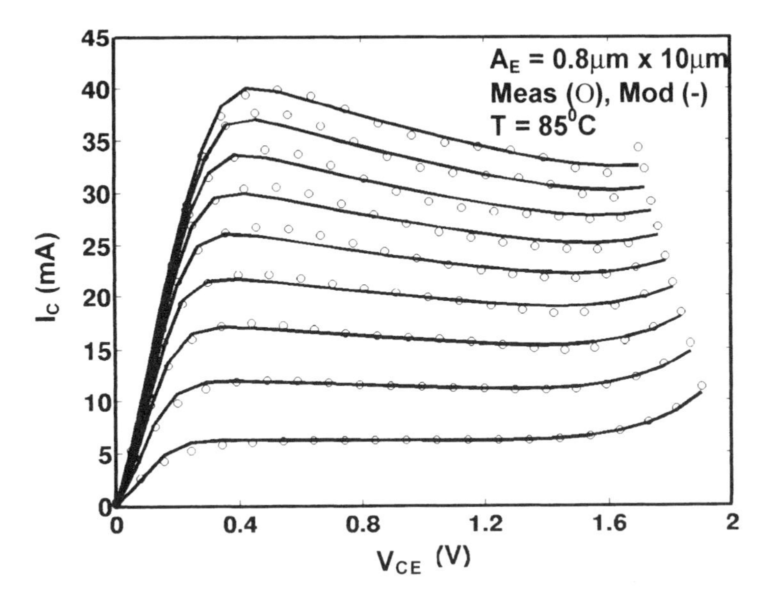

Figure 3.24. Measured (o) versus modeled (-) DC I–V characteristics at 85°C; I_B= 25–225μA in 25-μA increments.

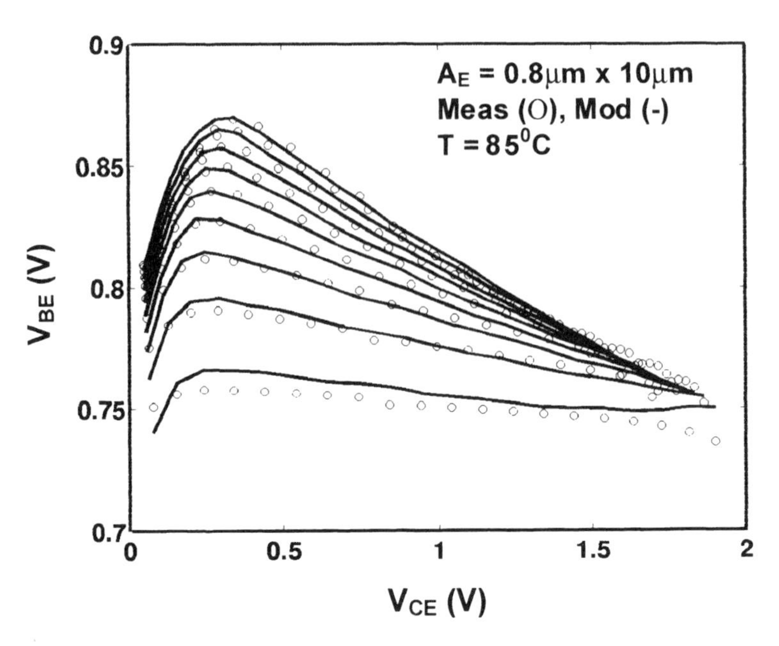

Figure 3.25. Measured (o) versus modeled (-) V_{BE}–V_{CE} characteristics at 85°C; I_B= 25–225 μA in 25-μA increments.

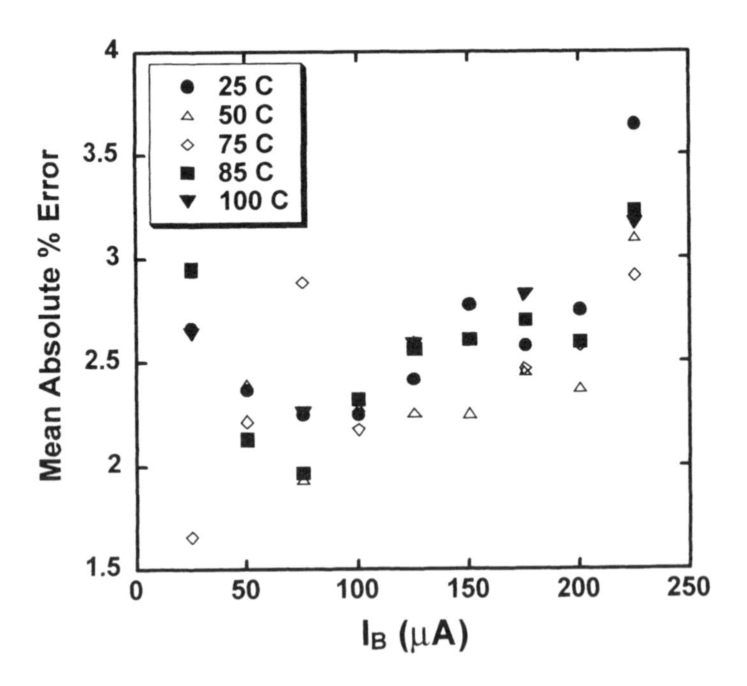

Figure 3.26. Average absolute percentage error versus I_B at different ambient temperatures.

Thus, the empirical modeling methodology can easily be adapted to model SiGe HBTs, making appropriate modifications in the model equations, according to the physics of the device. The bias-dependent intrinsic elements can be extracted from *S*-parameters measured at various biases and temperatures, similar to the case of the GaAs HBT and Si BJT models, to complete the large-signal model.

3.6 SUMMARY

The extensive model extraction and verification results presented in the chapter, for a variety of bipolar device technologies, establish the validity, flexibility, and usefulness of physics-based empirical modeling for the development of unified large-signal models of bipolar transistors. This modeling technique is attractive for building scalable models suitable for accurate prediction of the characteristics of the large power transistor cells used in RF power amplifier design. Given a layout configuration for building power cells, model parameters may be extracted with relatively little effort for a set of devices of varying sizes with the same layout construction, and empirical scaling equations can be formulated for the model parameters. Such a technique simplifies the task of building scalable models, removing the complexity in physically analyzing and formulating nonlinear parasitic and thermal scaling, and makes it feasible to build accurate models for RF power amplifier simulation and design.

REFERENCES

1. B. R. Ryun and I. M. Abdel-Motaleb, Effect of recombination current on current gain of HBTs, *IEE Proc.*, **138**(Pt. G): 1115–1119 (April 1991).
2. P. C. Grossman, and J. Choma, Large signal modeling of HBT including self-heating and transit time effects, *IEEE Trans. Microwave Theory Tech.*, **40**(3): 449–464 (March 1992).
3. K. Lu, P. Perry and T. Brazil, A new large signal AlGaAs/GaAs HBT model including self-heating effects, with corresponding parameter extraction procedure, *IEEE Trans. Microwave Theory Tech.*, **43**(7): 1433–1445 (July 1995).
4. A. Samelis and D. Pavlidis, A heterojunction bipolar transistor large-signal model for high power microwave applications, *IEEE MTT-S Int. Microwave Symp.Digest*, 1995, pp. 1231–1234.
5. M. Rudolph, R. Doerner, K. Beilenhoff, and P. Heymann, Scalable GaInP/GaAs HBT large-signal model, *IEEE Trans. Microwave Theory Tech.*, **48**(12): 2370–2376, (Dec. 2000).
6. M. S. Shirokov, S. V. Cherepko, D. Xiaohang, J. C. M. Hwang, and D. A. Teeter, Large-signal modeling and characterization of high-current effects in InGaP/GaAs HBTs, *IEEE Trans. Microwave Theory Tech.*, **50**(4): 1084–1094 (April 2002).

7. C.-J. Wei, J. C. M. Hwang, W .-J. Ho, and J. A. Higgins, Large signal modeling of self-heating, collector transit-time, and RF breakdown effects in power HBTs, *IEEE Trans. Microwave Theory Tech.*, **44**(12): 2641–2647 (Dec. 1996).
8. C. M. Snowden, Large-signal microwave characterization of AlGaAs/GaAs HBTs based on a physics-based electrothermal model, *IEEE Trans. Microwave Theory Tech.*, **45**(1): 58–71 (Jan. 1997).
9. L. H. Camnitz, S. Kofol, T. Low, and S. R. Bahl, An accurate, large-signal, high frequency model for GaAs HBTs, *Proc. IEEE GaAs IC Symp.*, 1996, pp. 303–306.
10. S. V. Cherepko, and J. C. M. Hwang, VBIC model applicability and extraction procedure for InGaP/GaAs HBTs, *Proc. Asia-Pacific Microwave Conf.*, 2001, pp. 716–721.
11. C.-J. Wei, J. Gering, S. Sprinkle, A. Klimahsow, and Y. A. Tkachenko, Modified VBIC-model for InGaP/GaAs HBTs, *Proc. Asia-Pacific Microwave Conf.*, 2002, pp. 354–357.
12. P. Baureis, D. Seitzer, and U. Schaper, Modeling of self-heating in GaAs/AlGaAs HBT's for accurate circuit and device analysis, *GaAs IC Symp. Digest*, 1991, pp. 125–128.
13. A. Ibarra and J. Garcia, Strategy for DC parameter extraction in bipolar transistors, *IEE Proc. G*, **137**: 5–11 (1990).
14. C.-J. Wei and J. C. M. Hwang, Direct extraction of equivalent circuit parameters for heterojunction bipolar transistors, *IEEE Trans. Microwave Theory Tech.*, **43**(9): 2035–2039 (Sept. 1995).
15. M. Rudolf, R. Doerner, and P. Heymann, Direct extraction of HBT equivalent circuit elements, *IEEE Trans. Microwave Theory Tech.*, **47**(1): 82–84 (Jan. 1999).
16. B. Sheinman, E. Wasige, M. Rudolph, R. Doerner, V. Sidorov, S. Cohen, and D. Ritter, A peeling algorithm for extraction of the HBT small-signal equivalent circuit, *IEEE Trans. Microwave Theory Tech*, **50**(12): 2804–2810 (Dec. 2002).
17. D. Teeter and W. R. Curtice, Comparison of hybrid pi and tee HBT topologies and their relationship to large signal modeling, *IEEE Int. Microwave Symp. (MTT-S) Digest*, 1997, pp. 375–378.
18. K. Lu, X. Zhang, and G. H. Henderson, A simplified large-signal HBT model for RF circuit design, *1998 IEEE Int. Microwave Symp. (MTT-S) Digest*, 1998, pp. 1607–1610.
19. A. Raghavan, B. Banerjee, S. Venkataraman, and J. Laskar, Direct extraction of InGaP/GaAs HBT large signal model, *2002 IEEE GaAs IC Symp. Digest*, Oct. 2002, pp. 225–228.
20. A. Raghavan, S. Venkataraman, B. Banerjee, D. Heo, and J. Laskar, Direct extraction of an empirical temperature-dependent InGaP/GaAs HBT large-signal model, *IEEE J. Solid-State Circuits*, **38**(9): 1443–1450 (Sept. 2003).
21. F. X. Sinnesbichler, and G. R. Olbrich, Accurate large-signal modeling of SiGe HBTs, *2000 IEEE Int. Microwave Symp. (MTT-S) Digest*, 2000, pp. 749–752.
22. Y. Suh, D. Heo, A. Raghavan, E. Gebara, S. Nuttinck, K. Lim, and J. Laskar, Direct extraction and modeling method for temperature dependent large signal CAD model of Si-BJT, *2001 IEEE MTT-S Int. Microwave Symp. Digest,* May 2001, pp. 971–974.
23. S. Bousnina, P. Mandeville, A. B. Kouki, R. Surridge, and F. M. Ghannouchi, Direct parameter extraction method for HBT small-signal model, *IEEE Trans. Microwave Theory Tech.*, **50**(2): 529–536 (Feb. 2002).
24. Y. Gobert, P. J. Tasker, and K. H. Bachem, A physical, yet simple, small-signal equivalent circuit for the heterojunction bipolar transistor, *IEEE Trans. Microwave Theory Tech*, **45**(1): 149–153 (Jan. 1997).

25. I. Angelov, K. Choumei, and A. Inoue, An empirical HBT large signal model for CAD, *2002 IEEE Int. Microwave Symp. (MTT-S) Digest*, June 2002, pp. 2137–2140.
26. W. J. Kloosterman, J. C. J. Paasschens, and R. J. Havens, A comprehensive bipolar avalanche multiplication compact model for circuit simulation, *IEEE Bipolar/BiCMOS Circuits Technology Meeting*, 2000, pp. 172–175.

4

SCALABLE MODELING OF RF MOSFETS

4.1 INTRODUCTION

In this chapter, we discuss important issues in the modeling of MOSFETs for RF power amplifier design, including the geometry scaling of critical parameters and their implementation in a large-signal MOSFET model. In particular, transistor modeling for RF power amplifier design needs to facto in the effects of external parasitics, as these parameters are not negligible in the large devices used in high-power amplifiers. For instance, if the substrate resistance in a bulk-CMOS process is not properly modeled, the simulated performance of the power amplifier can be significantly different from the actual characteristics. Gate resistance is another important element, because of its strong effect on input impedance, thermal noise, and the effective gain–bandwidth product of the transistor. To illustrate, BSIM3 accurately simulates digital and analog circuits, but typically produces large errors in S-parameter prediction. The reasons become clear when we realize that S_{11}, or the return loss, is highly dependent on the input resistance, and that S_{22}, the output reflection coefficient, would be inaccurate if the substrate network is not properly modeled. In this chapter, modifications to the standard BSIM3v3 model to make it more suitable for RF IC simulation are discussed in detail. BSIM3 was introduced in Chapter 2. It is an industry-standard MOSFET model for deep-submicrometer applications. The BSIM3 version 3 model (BSIM3v3), introduced in 1996, has been in use worldwide for CMOS IC design. It has been validated to be very accurate for digital applications at several hundred megahertz. However, its predictive performance in the RF and microwave frequency regime leaves much to be desired. Recent studies have shown that the MOSFET behaves as a distributed device and consequently, MOSFET models must be modified to reflect this, in

RF applications. In particular, modifications should be made to incorporate the following effects: distributed channel or the nonquasistatic (NQS) effect, distributed gate resistance, and distributed substrate resistance. Modeling these effects is important to accurately predict quantities such as power gain, input and output impedance, and the phase delay between drain current and gate voltage, which are important in RF IC design.

4.1.1 NQS Effects

An understanding of quasi-static operation in the MOS transistor under large-signal dynamic operation is important for analyzing NQS effects. In DC operation, a MOS transistor is driven by four DC voltages, V_D, V_G, V_B, and V_S, that is, the drain, gate, bulk and source voltages, respectively, defined with respect to ground. The channel of the device is formed by the electrons flowing from source to drain (assuming an NMOS device) in the inversion layer. Here, the device is assumed to be operating quasistatically, that is, the charge per unit area in the inversion layer and the depletion region are unchanged. This is valid when the variation of terminal voltages is sufficiently slow, that is, at DC and low frequency. Hence, for low-frequency operation, a model developed under the quasistatic assumption can be utilized with negligible error. However, when terminal voltages change more rapidly, as they do in high-frequency operation, the charges flowing in the inversion layer do not have sufficient time to follow the terminal voltages and the quasistatic assumption is no longer valid. Therefore, for high-frequency applications, the model must be developed specifically by considering finite channel transit times, or the so-called nonquasistatic operation.

The existence and impact of NQS effects on high-frequency MOSFET operation is well known [1,2]. There are several ways to account for NQS effects in MOSFET modeling. One method is to consider these effects using a solution of the current continuity equation [2–9]. Essentially, in this method, the channel is sectioned into many small pieces, and each section is solved for in a quasistatic manner. However, this approach requires extensive mathematical computation and is relatively complex compared to other methods. Also, this method is valid mostly for long-channel MOSFET devices and is not accurate for short-channel devices. An easier approach to model NQS effects, by using a single resistor connecting to the gate of the device, was proposed by Tin et al. [7]. This resistor introduces a delay to the charge traveling in the channel [10,11]. This method is more attractive since it models not only NQS effects but also, simultaneously, the gate-induced thermal noise. In BSIM3v3, NQS effects are modeled and can be selected by setting the model parameter *NQSMOD* to 1. However, this is available only for BSIM3v3 version 3.2 or later.

4.1.2 Distributed Gate Resistance

The effect of gate resistance is significant in the microwave frequency range. High gate resistance increases thermal noise and reduces the maximum available gain. In a deep-submicrometer salicided CMOS process, the polysilicon gate is silicided, to dramatically reduce the gate sheet resistance, typically to less than 10 Ω per square. Still, accurate modeling of gate resistance is important for high-frequency circuit simulation. The total gate resistance depends on the physical structure, that is, the layout, of the device. At high frequency, gate poly can be regraded as a distributed R–C circuit, and the gate resistance can be modeled as a lumped element using transmission-line theory, as given by [12]

$$R_g = \left(\frac{1}{3}\right) R_{square} \left(\frac{W}{L}\right) \tag{4.1}$$

where R_{square} is the gate sheet resistance and W and L are the gate finger width and length, respectively. This equation assumes that the gate is connected from one side. If both sides of gate poly are connected, the factor 1/3 will become 1/12.

This calculation, however, requires further adjustment for interconnections between different gate poly stripes in multifinger devices, which have not yet been accounted for. Furthermore, this calculation should be verified with the extracted value from measurement, to ensure correctness. For BSIM3v3, the gate resistance must be added externally to increase model accuracy at microwave frequencies. This gate resistance can be lumped with the resistance accounting for the NQS effects mentioned earlier, thus reducing the number of components.

4.1.3 Distributed Substrate Resistance

Substrate resistance has significant impact on high-frequency device performance. In RF CMOS IC design, the nonepitaxial (nonepi) process is preferred to the epitaxial (epi) process because of its higher substrate resistance. In a nonepi CMOS process, a high substrate resistance (6 to 10 Ω·cm) helps reduce the parasitic capacitance to the substrate and unwanted signal coupling between devices. The substrate resistance of the highly doped epi process is very small (0.01–0.02 Ω·cm), which helps protect the circuit from latchup. In this chapter, we discuss the modeling of substrate resistance for N-MOSFETs in a nonepi 0.18-μm CMOS process. It should be noted, however, that there is no difference in the modeling of substrate resistance between the two types of processes.

Substrate resistance can be calculated using a 2D device simulator, such as, MEDICI [13]. However, in a real device, substrate coupling effects are distributed in every direction in the p substrate, and thus cannot be captured easily by any kind of physical device simulator. Therefore, direct extraction from S-parameters (or Y-parameters) is usually the best method of obtaining accurate substrate parameters. In theory, substrate coupling effects can be modeled as a network of capacitors and resistors. The capacitors model the depletion capacitance of the drain–substrate or source–substrate junction, and the resistors represent the substrate resistance. In the BSIM3v3 model, a diode connected in reverse fashion is used, instead of a fixed capacitor, to model the bias dependence of the depletion capacitor. However, BSIM3v3 does not include substrate resistance. As a result, addition of external resistors from the internal bulk terminal to the external bulk pin is required, as shown in Figure 4.1.

The silicon substrate can be modeled as a single resistor for a frequency of up to 10 GHz, and as a parallel R–C circuit for higher frequency [14–16]. The main advantage of the single-resistor network over the parallel R–C network is the resulting simplicity in parameter extraction. Substrate modeling can be very complicated, or even impossible, if too many unknown parameters exist. It has been proved [14–16] that a resistive network is adequate for modeling the silicon substrate up to 20 GHz with negligible errors.

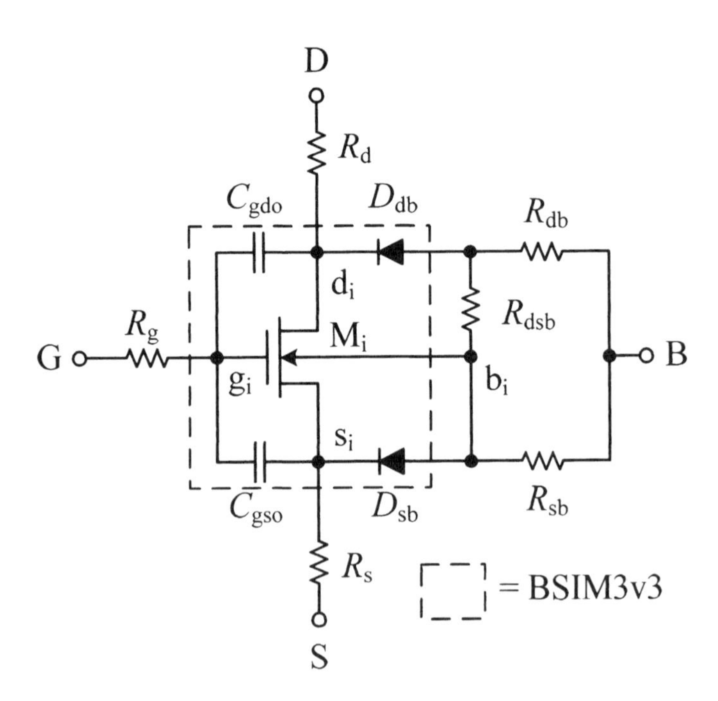

Figure 4.1. Modified BSIM3v3 model with substrate resistance network added.

The degree of substrate coupling strongly depends on the device geometry, that is, the distance from the source and drain junctions to the bulk connection. The bulk connection in RF MOS circuits is usually placed so as to minimize back-gate modulation and the possibility of latchup. In both cases, the shortest path between the intrinsic device and substrate connection with minimum substrate resistance is preferred. Therefore, the device is usually surrounded by a ring of substrate contact (sometimes called *guard ring* or *guard band*) with as many substrate connections as possible. However, most studies of substrate parameter extraction consider only side substrate contacts [7,14–22]. This is mainly because of the complexity in the 3D extraction of distributed parameters. However, such an approach limits accuracy in actual circuit simulation, where devices are always surrounded by guard rings. Figure 4.2 illustrates the two device geometries: with side substrate contacts alone, and the ring substrate contact structure. In this chapter, the development of a scalable MOSFET model including a substrate network, where the transistor is completely surrounded by substrate contacts, is described. The modeling methodology and step-by-step parameter extraction procedure are discussed in detail. Most importantly, the scalability of each parameter, which is key in developing a scalable large-signal model for the MOSFET, is explored and implemented.

4.2 SCALABLE MODIFIED BSIM3V3 MODEL

4.2.1 Scalability of MOSFET Model

Scalability of model parameters is one of the most important considerations for any transistor model. In the MOSFET, scalability depends on the geometry of the device [23,24]. The discussion here pertains to the scalability of each parameter of the small-signal equivalent circuit shown in Figure 4.3. Normally, intrinsic parameters, including the gate-source capacitance (C_{gs}), gate–drain capacitance (C_{gd}), small-signal transconductance (g_m), drain–source resistance (r_{ds}), and R_i, the internal resistance to model NQS effects [7], are linearly scalable with gate geometry (total gate width and width of each gate finger) since they are proportional to the underlying channel area. R_g, R_d, and R_s, which are extrinsic parameters (not part of the channel), also linearly scale with gate geometry. The substrate parameters, C_{sub} and R_{sub}, however, do not scale linearly with gate geometry because of their distributed behavior, scattering in every direction from the n+ drain and source (for an N-MOSFET) to the bulk contacts located on the side and/or top and bottom of the device. Therefore, the modeling of the substrate coupling parameters is more complicated than the other parameters.

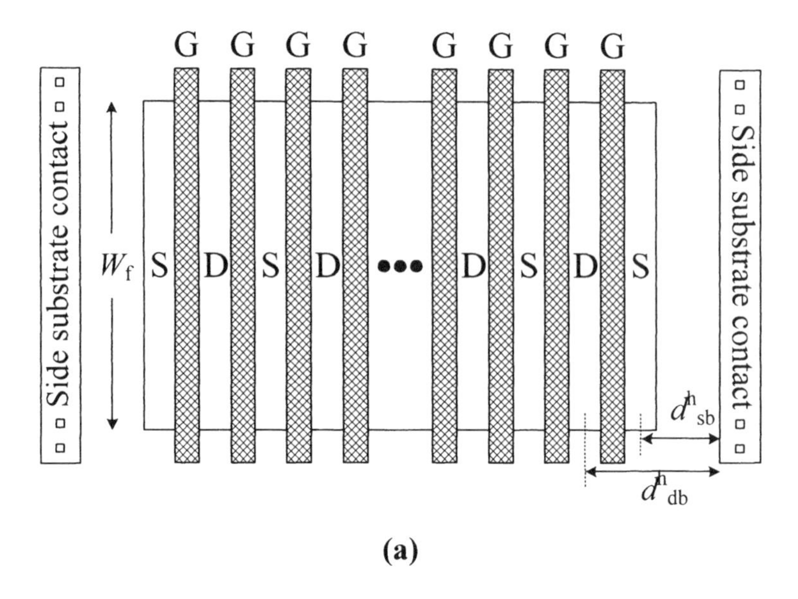

(a)

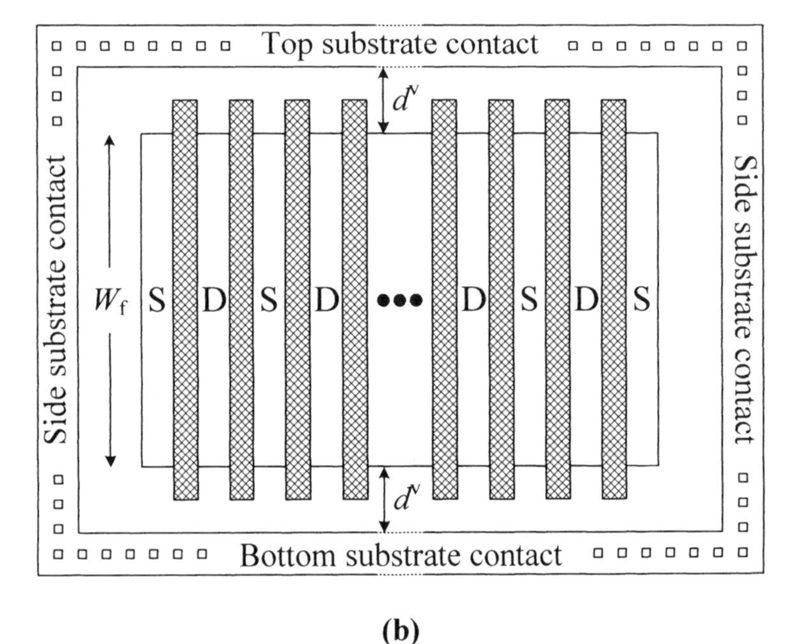

(b)

Figure 4.2. MOSFET device layout configurations used in substrate modeling considering (a) only side substrate contacts and (b) ring-shaped contacts (d^h and d^v represent the distance from the active junctions to the substrate contact in the horizontal and vertical directions, respectively).

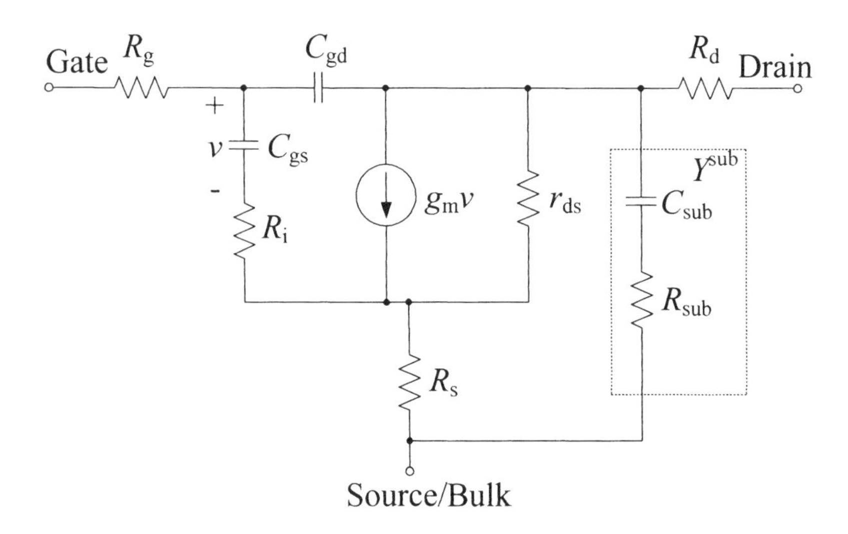

Figure 4.3. Small-signal MOSFET equivalent circuit including the substrate network (labeled Y^{sub}).

Most published scalable MOSFET substrate models [14,17,19] are limited to a relatively small range of device periphery (up to 500 μm total gate width) and consider only substrate coupling from drain and source junctions to the side body contacts (as in Figure 4.2a) due to the increased complexity in parameter extraction of larger devices. However, large devices (up to several millimeters) are required in many applications such as driver amplifiers and power amplifiers, where accuracy of the parameter extraction and scalability are crucial. Moreover, the ring-shaped substrate contact (see Figure 4.2b) is commonly used for reducing body effect in most RF circuit designs, and modeling only the coupling form drain and source to the side body contacts results in an inaccurate prediction of circuit performance. It is found that the discrepancy in parameter scalability using this model [14,19] and a model with ring-shaped substrate contact is due to the fact that the substrate coupling from the drain and source junctions to the top and bottom substrate contacts was neglected in the former. This additional substrate coupling is significant, yet difficult to model with compact analytical expressions, because of its distributed nature throughout the device. Empirical expressions can be used to model substrate parameter scalability [25]. However, the use of empirical equations limits model scalability to certain device geometries used in the modeling process and makes it difficult to extend the study for different device structures.

The scalable model discussed in this chapter utilizes analytical expressions for substrate parameter modeling based on device geometry, that is, the number of gate fingers (N_f), gate finger width (W_f) and length (L_f), and parameters related to the geometry of the ring-shaped substrate contact. The model considers the effect of substrate coupling to the entire ring-shaped substrate contact and approximates it to substrate coupling in the vertical and horizontal directions. This method accurately models the scalability of substrate parameters for a large range of device sizes up to 6 mm total gate width. The parameter extraction procedure and the modeling of parameter scaling are presented in the following sections.

4.2.2 Extraction of Small-Signal Model Parameters

Comparing the small-signal model (Figure 4.3) and the BSIM3v3 model (Figure 4.1), the small-signal model is actually a reduced form of BSIM3v3 when the source and bulk are tied together. Therefore, the extraction of substrate resistance and gate resistance for BSIM3v3 can be done by first extracting the substrate resistance and gate resistance using the small-signal equivalent model. Then, the parameters are incorporated in the BSIM3v3 model and further optimized to fit the measurement results.

The small-signal model parameters of Figure 4.3 can be extracted by analyzing the two-port *Y*-parameters of the device. First, the *S*-parameters of N-MOSFET test devices, with the total gate width ranging from 200 μm to 6 mm, are measured [using a vector network analyzer (VNA), such as the HP8510]. The MOSFETs, which have unit finger width of 20 μm, were fabricated in a standard five-metal layer CMOS process with a gate length of 0.4 μm. As commonly used in RF applications, the substrate contact is formed as a ring surrounding the active device area, as illustrated in Figure 4.2b, to minimize the back-gate effect. The substrate contact is tied to the source terminal, which is common practice in RF circuit design. Multiple gate fingers of the same size were used to construct large FETs without changing the layout style, in order to minimize the discrepancy in the parasitics associated with the layout. Figures 4.4 and 4.5 show photographs of the test devices used.

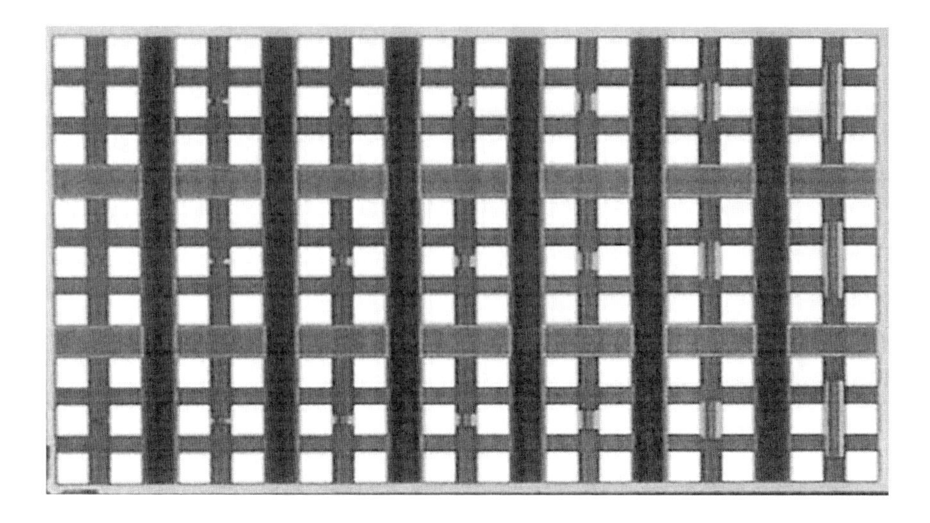

Figure 4.4. Photograph of NMOSDG test structures of total gate width ranging from 200 μm to 6 mm. The first and second rows contain short and open structures for deembedding pad and interconnect parasitics. The third row contains the actual MOSFET devices.

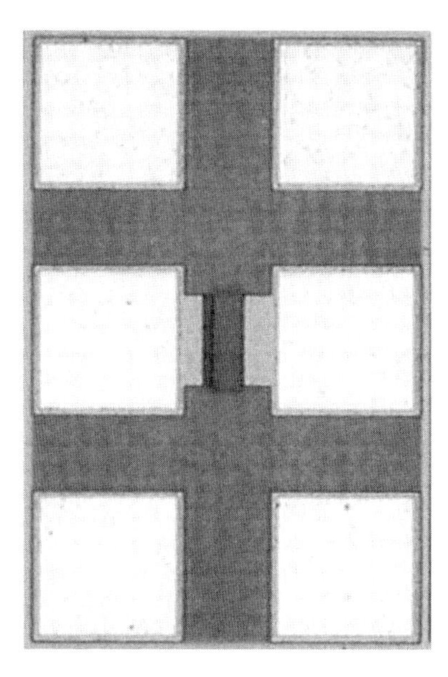

Figure 4.5. Photograph of an NMOSDG test structure. The MOSFET device, with a total gate width of 1200 μm in this test structure, is at the center. The two sets of three pads each, on the left and right, are in a ground–signal–ground configuration (top to bottom in each case), and serve as landing sites for the microwave probes used for measuring the *S*-parameters.

Model parameter extraction is similar to the method described by Cheng and Matloubian [14,17], and high-frequency measurement data up to 40 GHz are used to find parasitic resistances R_g, R_d, and R_s. Before the intrinsic parameters can be calculated from the measured Y-parameters, the Y-parameters of the substrate network have to be calculated and subtracted out. At zero-bias ($V_g = 0$ V, $V_{ds} = 0$ V), substrate parameters can be extracted separately from the intrinsic parameters. The substrate parameter extraction is done using following expressions for R_{sub} and C_{sub}

$$R_{sub} = \frac{\mathrm{Re}\{Y_{22}^{c}\}}{\left(\mathrm{Im}\{Y_{22}^{c}\} + \mathrm{Im}\{Y_{12}^{c}\}\right)^2} \tag{4.2}$$

$$C_{sub} = \frac{\mathrm{Im}\{Y_{22}^{c}\} + \mathrm{Im}\{Y_{12}^{c}\}}{\omega} \tag{4.3}$$

where Y^c are the Y-parameters after deembedding the pad and interconnect parasitics, as well as removing the gate resistance (R_g) and drain contact resistance (R_d). R_{sub} is bias-independent, and can be readily included in the equivalent small-signal model at the desired bias condition [17]. The schematic diagram of pad parasitics and interconnects is shown in Figure 4.6, and the equations for finding pad and interconnect parasitics are as follows:

$$C_{pgd} = -\frac{1}{\omega}\left(\frac{\mathrm{Re}\{Y_{12}^{o}\}^2}{\mathrm{Im}\{Y_{12}^{o}\}} + \mathrm{Im}\{Y_{12}^{o}\}\right) \tag{4.4}$$

$$C_{pgs} = \frac{1}{\omega}\left(\frac{\mathrm{Re}\{Y_{11}^{o} + Y_{12}^{o}\}^2}{\mathrm{Im}\{Y_{11}^{o} + Y_{12}^{o}\}} + \mathrm{Im}\{Y_{11}^{o} + Y_{12}^{o}\}\right) \tag{4.5}$$

$$C_{pds} = \frac{1}{\omega}\left(\frac{\mathrm{Re}\{Y_{22}^{o} + Y_{12}^{o}\}^2}{\mathrm{Im}\{Y_{22}^{o} + Y_{12}^{o}\}} + \mathrm{Im}\{Y_{22}^{o} + Y_{12}^{o}\}\right) \tag{4.6}$$

$$R_{pgd} = \frac{1}{\omega C_{pgd}}\sqrt{\frac{-\omega C_{pgd}}{\mathrm{Im}\{Y_{12}^{o} - 1\}}} \tag{4.7}$$

$$R_{pgs} = \frac{1}{\omega C_{pgs}}\sqrt{\frac{\omega C_{pgs}}{\mathrm{Im}\{Y_{11}^{o} + Y_{12}^{o} - 1\}}} \tag{4.8}$$

$$R_{\text{pds}} = \frac{1}{\omega C_{\text{pds}}} \sqrt{\frac{\omega C_{\text{pds}}}{\text{Im}\{Y_{22}^{\text{o}} + Y_{12}^{\text{o}} - 1\}}} \tag{4.9}$$

$$R_{\text{sgg}} = \text{Re}\{Z_{11}^{\text{s}} - Z_{12}^{\text{s}}\} \tag{4.10}$$

$$R_{\text{sdd}} = \text{Re}\{Z_{22}^{\text{s}} - Z_{12}^{\text{s}}\} \tag{4.11}$$

$$R_{\text{sss}} = \text{Re}\{Z_{12}^{\text{s}}\} \tag{4.12}$$

$$L_{\text{sgg}} = \frac{\text{Im}\{Z_{11}^{\text{s}} - Z_{12}^{\text{s}}\}}{\omega} \tag{4.13}$$

$$L_{\text{sdd}} = \frac{\text{Im}\{Z_{22}^{\text{s}} - Z_{12}^{\text{s}}\}}{\omega} \tag{4.14}$$

$$L_{\text{sss}} = \frac{\text{Im}\{Z_{12}^{\text{s}}\}}{\omega} \tag{4.15}$$

where C_{pgd}, C_{pgs}, C_{pds}, R_{pgd}, R_{pgs}, and R_{pds} are the parasitic capacitances and resistances associated with each pad and R_{sgg}, R_{sdd}, R_{sss}, L_{sgg}, L_{sdd}, and L_{sss} are interconnect parasitic resistances and inductances (these are excluded from the intrinsic device). The removal of pad and interconnect parasitics must be done prior to the extraction of intrinsic device parameters. To remove these parasitics, the following procedure is followed:

$$Y_{\text{DUT,OPEN}} = Y_{\text{DUT}} - Y_{\text{OPEN}} \tag{4.16}$$

$$Y_{\text{SHORT,OPEN}} = Y_{\text{SHORT}} - Y_{\text{OPEN}} \tag{4.17}$$

$$Z_{\text{DUT,OPEN,SHORT}} = Z_{\text{DUT,OPEN}} - Z_{\text{SHORT,OPEN}} \tag{4.18}$$

where $Y_{\text{DUT,OPEN}}$ are the Y-parameters of the device after the deembedding the pad parasitics using the "open" test structure (structure with transistor removed), $Y_{\text{SHORT,OPEN}}$ are the Y-parameters of the "short" test structure (where the transistor is removed and a small piece of metal to short all paths is inserted) after deembedding the pad parasitics, and $Z_{\text{DUT,OPEN,SHORT}}$ are the Z-parameters of the intrinsic device after removing all the pad and interconnect parasitics. Details of the parasitic components are shown in Figure 4.7. Equations for transforming between Z, Y, and S parameters can be found in Pozar's text [26].

Next, the gate, drain, and source resistances (R_g, R_d, and R_s) are extracted using the cold-FET method, where the measurements are made at zero bias, using the following equations:

$$R_g = \mathrm{Re}\left\{Z_{11}^{\mathrm{cold}} - Z_{12}^{\mathrm{cold}}\right\} \tag{4.19}$$

$$R_d = \mathrm{Re}\left\{Z_{22}^{\mathrm{cold}} - Z_{12}^{\mathrm{cold}}\right\} \tag{4.20}$$

$$R_s = \mathrm{Re}\left\{Z_{12}^{\mathrm{cold}}\right\} \tag{4.21}$$

The extraction is done assuming that the impedance of intrinsic capacitances at very high frequencies (> 20 GHz, for C_{gs} and C_{gd} < 100 fF) is negligible, shorting all capacitors and leaving only R_g, R_d, and R_s.

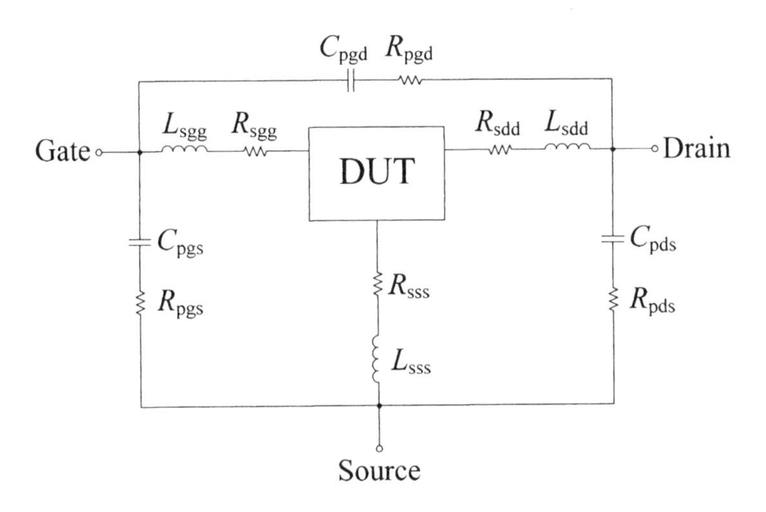

Figure 4.6. Pad and interconnect parasitics for the MOSFET test structures.

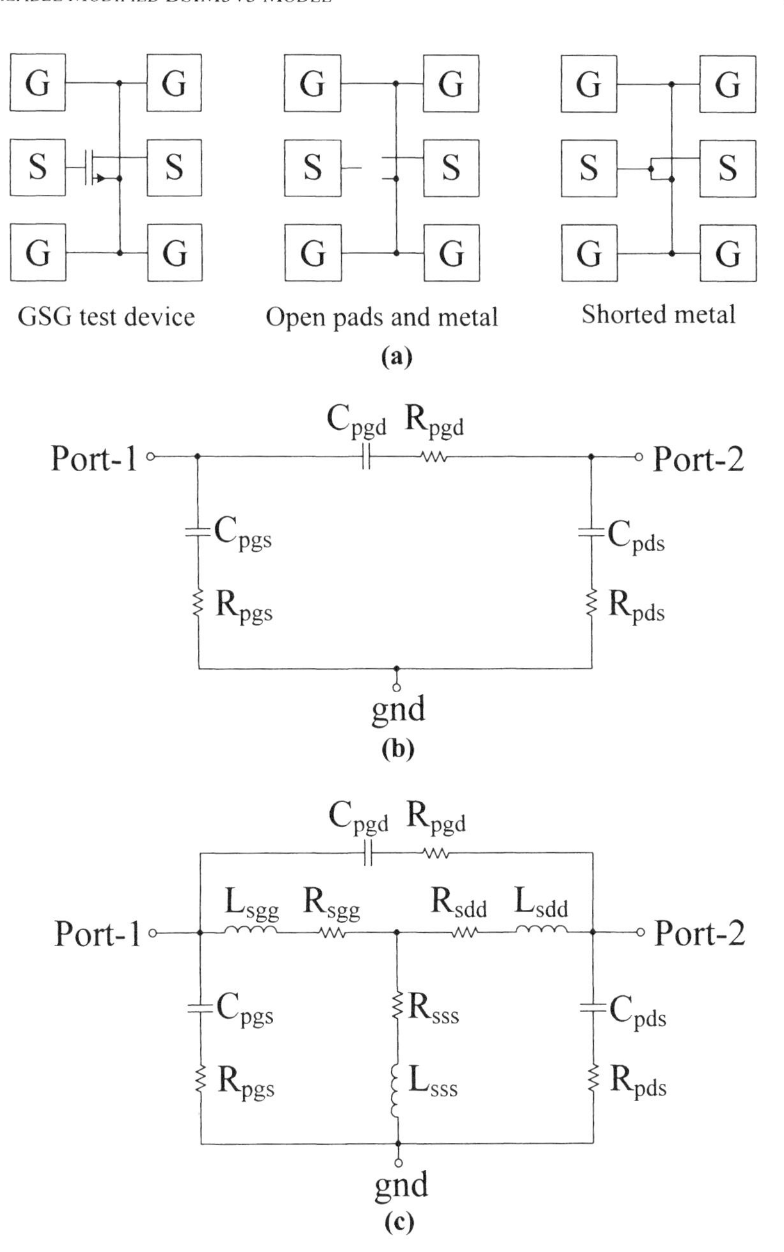

Figure 4.7. Test structures required for two-step removal of pad and interconnect parasitics: (a) test device, open and short test structures; (b) parasitic components for the open structure; (c) parasitic components for the short structure.

For higher-frequency applications (>20 GHz), the magnitude of the impedance corresponding to the terminal inductances (L_g, L_d, and L_s) may be comparable to the terminal resistances, and should be included for better accuracy. These inductances can be calculated using

$$L_g = \frac{\mathrm{Im}\left\{Z_{11}^{\mathrm{cold}} - Z_{12}^{\mathrm{cold}}\right\}}{\omega} \tag{4.22}$$

$$L_d = \frac{\mathrm{Im}\left\{Z_{22}^{\mathrm{cold}} - Z_{12}^{\mathrm{cold}}\right\}}{\omega} \tag{4.23}$$

$$L_s = \frac{\mathrm{Im}\left\{Z_{12}^{\mathrm{cold}}\right\}}{\omega} \tag{4.24}$$

The terminal resistances and inductances are removed before the extraction of R_{sub} and C_{sub} using Equations 4.2 and 4.3. To complete the intrinsic parameter extraction, R_{sub} and C_{sub} are first removed from the Y-parameters of the device after deembedding terminal resistances and inductances using Equations 4.19–4.24. Finally, the remaining intrinsic parameters are extracted using the following equations

$$C_{gd} = -\frac{\mathrm{Im}\left\{Y_{22}^{\mathrm{d}}\right\}}{\omega} \tag{4.25}$$

$$C_{gs} = \frac{\mathrm{Im}\left\{Y_{11}^{\mathrm{d}} + Y_{12}^{\mathrm{d}}\right\}}{\omega} \tag{4.26}$$

$$R_i = -\frac{\mathrm{Re}\left\{Y_{11}^{\mathrm{d}}\right\}}{\left(\omega C_{gs}\right)^2} \tag{4.27}$$

$$g_m = \mathrm{Re}\left\{Y_{21}^{\mathrm{d}}\right\} \tag{4.28}$$

$$\tau = -\frac{g_m \,\mathrm{Im}\left\{Y_{21}^{\mathrm{d}} - Y_{12}^{\mathrm{d}}\right\}}{\omega} - R_i C_{gs} \tag{4.29}$$

$$r_{ds} = \frac{1}{\mathrm{Re}\left\{Y_{22}^{\mathrm{d}}\right\}} \tag{4.30}$$

where Y^{d} are the Y-parameters of the device after deembedding the terminal resistances and inductances, and the substrate network C_{gd} is the gate–drain capacitance, C_{gs} is the gate–source capacitance, R_i is the internal resistance of the gate–source junction, g_m is the transconductance, τ is the channel time delay, and r_{ds} is the output resistance.

These parameters are extracted for a number of devices in several dies. Then, the most consistent extracted values are carefully selected without performing optimization in order to minimize measurement uncertainties. Most of the extracted parameters show a linear scaling ability. This includes the parasitic resistances (R_g, R_d, and R_s), gate–source capacitance (C_{gs}), gate–drain capacitance (C_{gd}), output resistance (r_{ds}), R_i, and transconductance (g_m). R_g, R_d, R_s, R_i, and r_{ds} scale in inverse proportion to W as they are dependent on the number of fingers connected in parallel (see Table 4.1 for values) C_{gs}, C_{gd}, and g_m are observed to be the parameters least sensitive to extraction uncertainties and also show good linear scalability (see Figure 4.8).

4.2.3 Scalable Substrate Network Modeling

The analysis of the substrate network discussed here is based on a set of 0.4-μm multifinger NMOS thick-oxide devices having gate finger width of 20 μm with ring-shaped body contacts surrounding the active device. This type of substrate connection is used widely in RF CMOS circuit design since it minimizes the voltage discrepancy between source junction and substrate. However, it increases the difficulty in developing an accurate scalable substrate model, since the distributed nature of substrate coupling is not limited to any one direction. In this section, a simplified model of substrate resistance is used, which considers two main paths – the substrate resistance from the junction to the side substrate contacts (horizontal), represented by R_{db}^{h} and R_{sb}^{h}; and to the top and bottom substrate contacts (vertical), modeled by R_{db}^{v} and R_{sb}^{v} (see Figure 4.9) – instead of considering the distribution to the entire substrate contact. This approximation is acceptable because the smaller resistance, which could be one of the two directions as stated, dominates the larger resistance in other directions. By this simplification, the equations for approximating substrate resistance can be formulated as follows

$$\frac{1}{R_{db}^{t}} = \frac{1}{R_{db}^{h}} + \frac{1}{R_{db}^{v}} = \sum_{k=1}^{N_d} \left(\frac{1}{R_{db,k}^{h}} + \frac{1}{R_{db,k}^{v}} \right) \tag{4.31}$$

$$\frac{1}{R_{sb}^{t}} = \frac{1}{R_{sb}^{h}} + \frac{1}{R_{sb}^{v}} = \sum_{k=1}^{N_s} \left(\frac{1}{R_{sb,k}^{h}} + \frac{1}{R_{sb,k}^{v}} \right) \tag{4.32}$$

$$\frac{1}{R_{dsb}^{t}} = \sum_{k=1}^{N_f} \frac{1}{R_{dsb,k}} \tag{4.33}$$

where $R_{\mathrm{db}}^{\mathrm{t}}$ and $R_{\mathrm{sb}}^{\mathrm{t}}$ are the total drain-to-bulk and total source-to-bulk resistances; $R_{\mathrm{dsb}}^{\mathrm{t}}$ is the equivalent resistance between drain and source underneath the channel in the substrate; and $R_{\mathrm{db,k}}^{\mathrm{h}}$, $R_{\mathrm{db,k}}^{\mathrm{v}}$, $R_{\mathrm{sb,k}}^{\mathrm{h}}$, $R_{\mathrm{sb,k}}^{\mathrm{v}}$, and $R_{\mathrm{dsb,k}}$ are the equivalent resistances corresponding to each drain and source junction in the horizontal (h) and vertical (v) directions. N_{f}, N_{d}, and N_{s} are the number of gate, drain, and source fingers, respectively. R_{sub} is an approximation of the substrate resistance from the substrate network equivalent circuit in Figure 4.10a.

$$R_{\mathrm{sub}} \approx \frac{R_{\mathrm{db}}^{\mathrm{t}}\left(R_{\mathrm{sb}}^{\mathrm{t}} + R_{\mathrm{dsb}}^{\mathrm{t}}\right)}{R_{\mathrm{db}}^{\mathrm{t}} + R_{\mathrm{sb}}^{\mathrm{t}} + R_{\mathrm{dsb}}^{\mathrm{t}}} \tag{4.34}$$

Table 4.1. Extracted intrinsic and substrate parameters ($V_{\mathrm{gs}} = 1.2$ V, $V_{\mathrm{ds}} = 2.4$ V)

Parameters	200 μm	800 μm	2400 μm
R_g (Ω)	12.7	3.3	1.3
R_d (Ω)	6.5	1.8	0.7
R_s (Ω)	7.8	1.9	0.6
R_i (Ω)	9.9	2.7	0.8
C_{gs} (fF)	246	993	2953
C_{gd} (fF)	55	227	703
g_m (mS)	42	170	513
R_{ds} (Ω)	856	204	63
C_{sub} (fF)	58	226	716
R_{sub} (Ω)	158	82	35

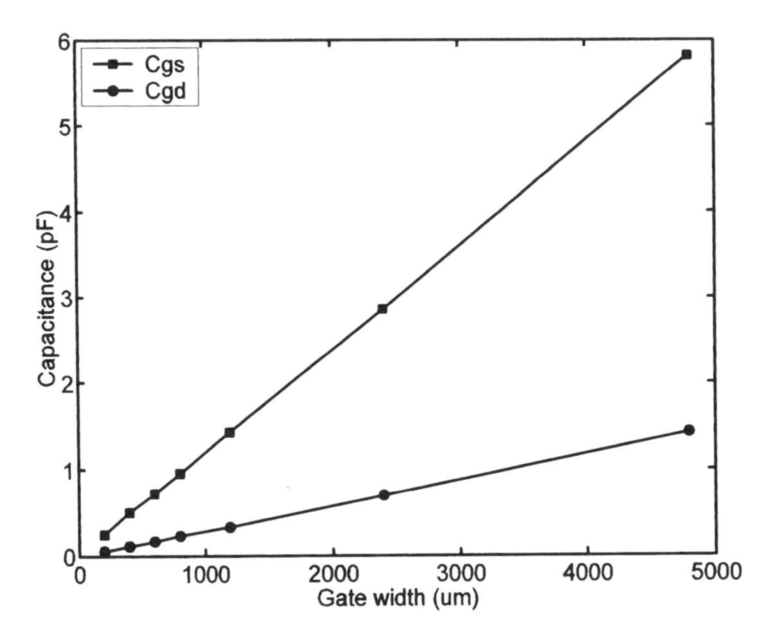

Figure 4.8. Scaling of C_{gs} and C_{gd} with gate width.

The total substrate capacitance can be calculated from the following equation

$$C_{sub} \approx \sum_{k=1}^{N_d} C_{db,k} \tag{4.35}$$

with the assumption that $(\omega C_{sb})^2 R_{sb}^2 << 1$ (valid for frequencies up to 10 GHz). The extraction of each substrate resistance parameter can be done using the extracted R_{sub} from several devices with different number of fingers. The contribution of the substrate resistance of the inner drain and source junctions to the side substrate contacts is very small compared to that from the sidemost junctions [14–18]. Therefore, R_{db}^h and R_{sb}^h can be approximated as contributions from the side most junctions to reduce extraction complexity. For a symmetric device layout (even number of gate fingers), R_{db}^h, R_{sb}^h, R_{db}^v, R_{sb}^v, and R_{dsb}^t can be calculated as

$$R_{db}^h \approx \frac{r_{db}^h}{2} \cdot \frac{d_{db}^h}{W_f} \tag{4.36}$$

$$R_{\mathrm{sb}}^{\mathrm{h}} \approx \frac{r_{\mathrm{sb}}^{\mathrm{h}}}{2} \cdot \frac{d_{\mathrm{sb}}^{\mathrm{h}}}{W_{\mathrm{f}}} \tag{4.37}$$

$$R_{\mathrm{db}}^{\mathrm{v}} \approx \alpha \cdot \frac{r_{\mathrm{db}}^{\mathrm{v}}}{2N_{\mathrm{d}}} \cdot \frac{d^{\mathrm{v}}}{L_{\mathrm{d}}} \tag{4.38}$$

$$R_{\mathrm{sb}}^{\mathrm{v}} \approx \alpha \cdot \frac{r_{\mathrm{sb}}^{\mathrm{v}}}{2N_{\mathrm{s}}} \cdot \frac{d^{\mathrm{v}}}{L_{\mathrm{s}}} \tag{4.39}$$

$$R_{\mathrm{dsb}}^{\mathrm{t}} \approx \frac{r_{\mathrm{dsb}}}{N_{\mathrm{f}}} \cdot \frac{L_{\mathrm{f}}}{W_{\mathrm{f}}} \tag{4.40}$$

where $r_{\mathrm{db}}^{\mathrm{h}}$, $r_{\mathrm{db}}^{\mathrm{v}}$, $r_{\mathrm{sb}}^{\mathrm{h}}$, $r_{\mathrm{sb}}^{\mathrm{v}}$, and r_{dsb} are unit substrate resistances in the horizontal and vertical directions corresponding to each junction; $d_{\mathrm{db}}^{\mathrm{h}}$ and $d_{\mathrm{sb}}^{\mathrm{h}}$ are the distances between the centers of the outmost drain and source fingers and the side substrate contact; d^{v} is the distance from the upper or lower edge of drain and source fingers to the top or bottom substrate contact; and L_{f}, L_{d}, and L_{s} are the device channel length, and the length of the drain and source junctions, respectively. The parameter α in Equations 4.38 and 4.39 is used to correct the distributed behavior of $R_{\mathrm{db}}^{\mathrm{v}}$ and $R_{\mathrm{sb}}^{\mathrm{v}}$ beneath the drain and source junctions coupled to the top and bottom substrate contacts ($0.5 \le \alpha \le 1$).

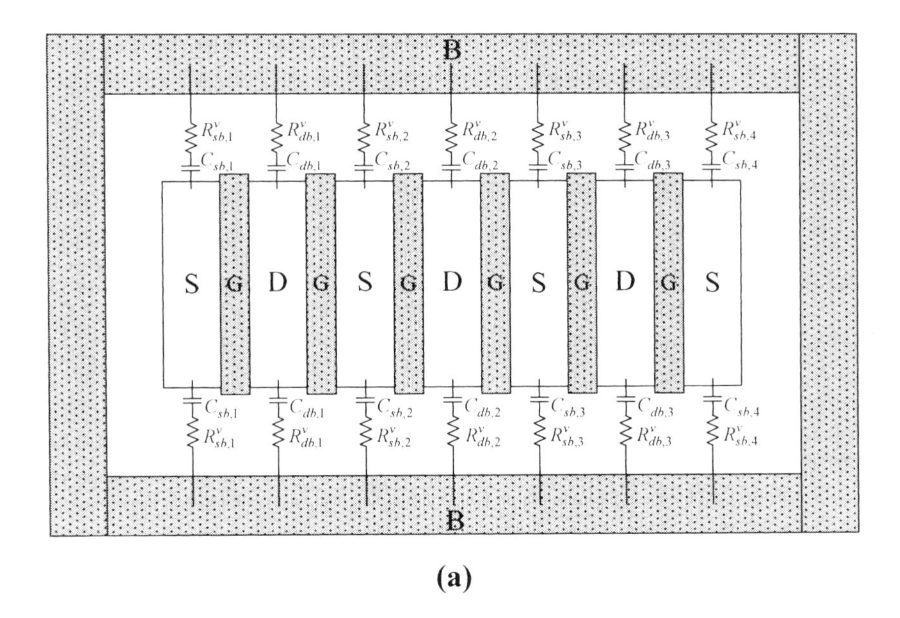

(a)

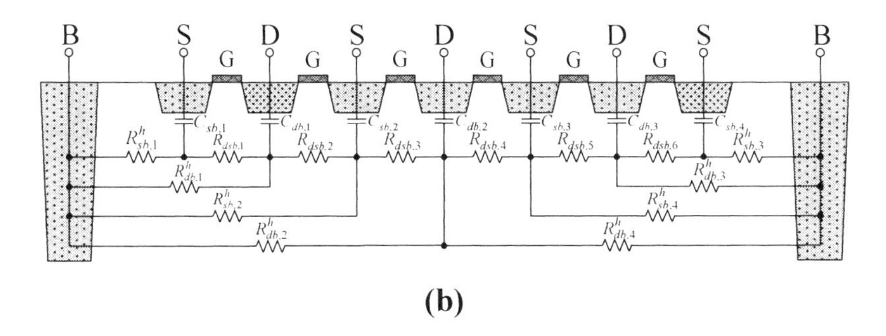

(b)

Figure 4.9. Substrate coupling in a MOSFET surrounded by body contacts: (a) vertical coupling; and (b) horizontal coupling.

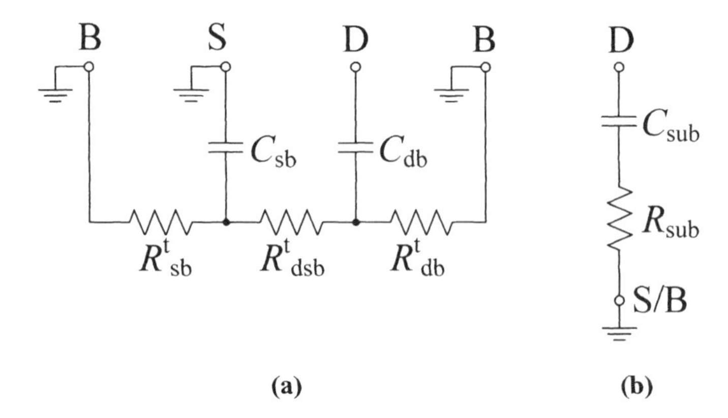

Figure 4.10. (a) Equivalent-circuit representation of the substrate parasitics; and (b) approximated substrate network.

The extracted substrate network parameters, R_{sub} and C_{sub}, of 0.4 μm multifinger NMOS thick-oxide devices with total gate width ranging from 200 μm to 6 mm are shown in Figure 4.11. All the devices use an identical gate finger width of 20 μm, $d^v = 2.8$ μm, $d^h_{db} = 6.6$ μm, $d^h_{sb} = 5.6$ μm, and $L_d = L_s = 0.6$ μm. The correction parameter α is obtained using Equations 4.36–4.40, assuming that the same unit substrate resistance is used in both directions. In this case, the parameter α is determined to be 0.5. It is clearly seen that this model can predict substrate parameters very accurately. Figure 4.12 shows the variation of R^t_{db}, R^t_{sb}, and R^t_{dsb} with device size for a given gate finger width. It is observed from Figure 4.13 that the substrate resistance components in the vertical direction (R^v_{db} and R^v_{sb}) are much smaller than those in the horizontal direction (R^h_{db} and R^h_{sb}), therefore dominating the R_{sub} value as the device size increases. This confirms the necessity of including the substrate coupling in the vertical direction in the modeling of substrate resistance.

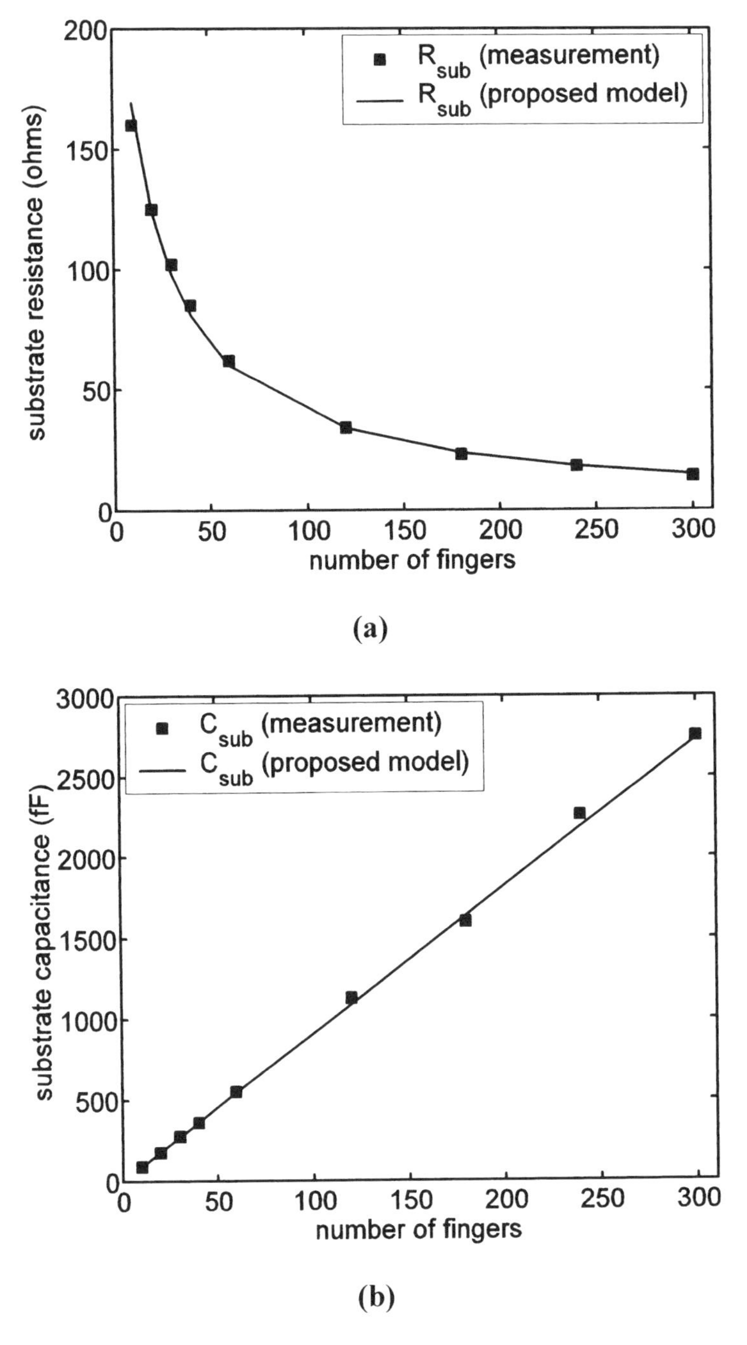

Figure 4.11. Substrate network parameter extraction results: (a) substrate resistance; and (b) substrate capacitance (V_{ds} = 0 V, V_{gs} = 0 V).

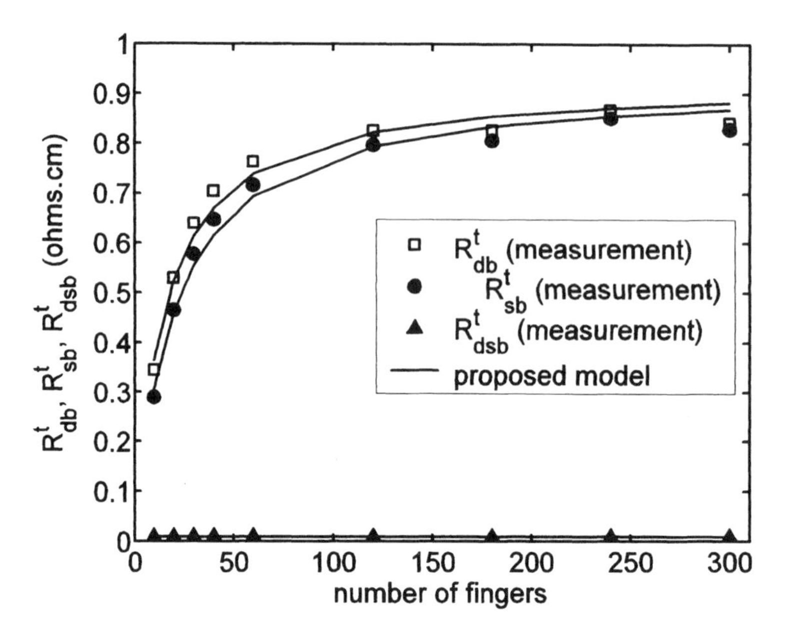

Figure 4.12. Scalability of substrate resistance components ($V_{ds} = 0$ V, $V_{gs} = 0$ V).

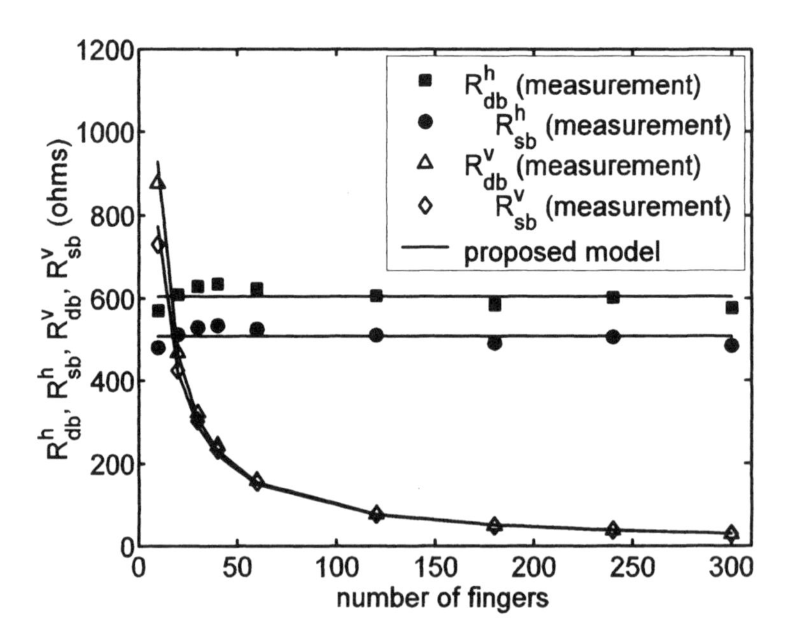

Figure 4.13. Substrate resistance components in the vertical and horizontal directions ($V_{ds} = 0$ V, $V_{gs} = 0$ V).

Substrate parameters are also extracted at various drain bias voltages to show their bias dependence. The results in Figure 4.14 show that the substrate resistance is bias-independent [14,17], while the substrate capacitance is bias dependent. The bias dependence of C_{db} (or C_{sb}) can be modeled by

$$C_{db} = \frac{C_{db0}}{\left(1 + \frac{V_{db}}{\varphi_{bi}}\right)^{\alpha_j}} \tag{4.41}$$

where C_{db0} is C_{db} at zero bias, φ_{bi} is the built-in potential, which is obtained by fitting measurement results to the model, and is found to be equal to 0.78 V here. α_j is a fitting parameter that is equal to 0.3 in this case. There is good agreement between the modeled characteristics and measurements, as seen from Figure 4.15.

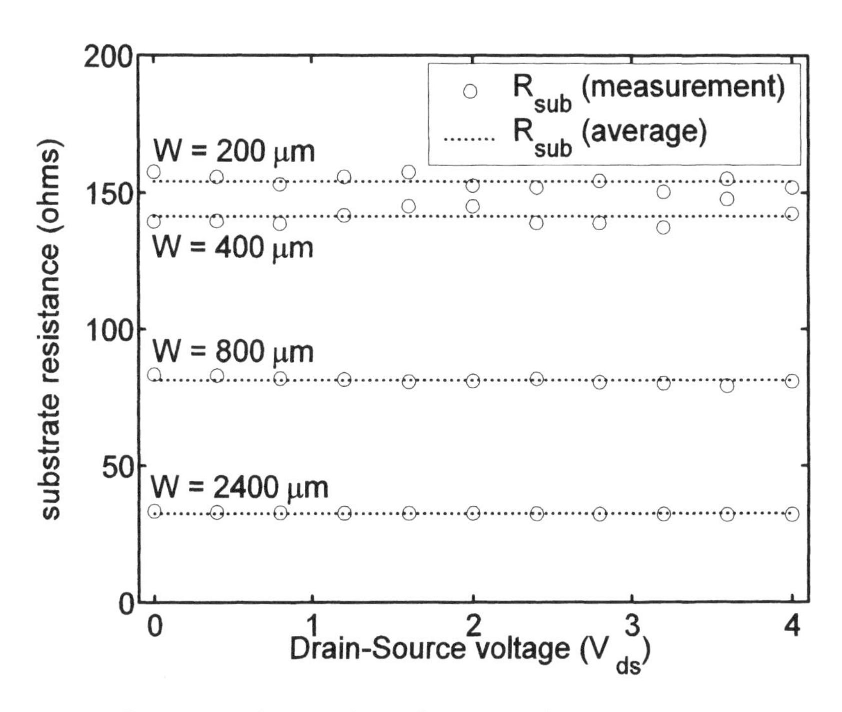

Figure 4.14. Bias dependence of substrate resistance (V_{gs} = 0 V).

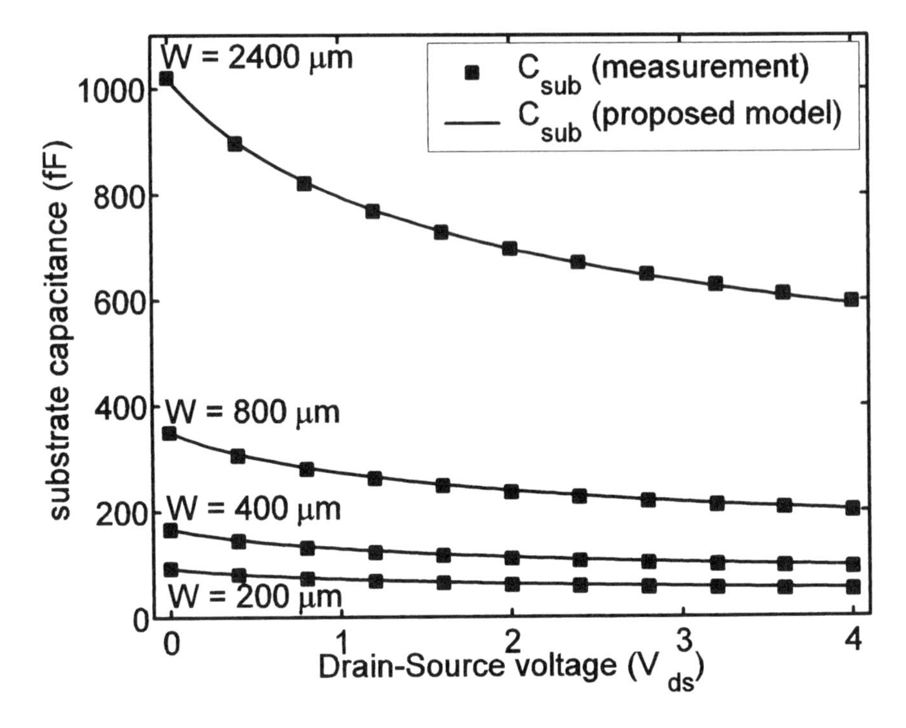

Figure 4.15. Bias dependence of substrate capacitance ($V_{gs} = 0$ V).

In addition to the previous results, substrate resistance and capacitance of 2400-μm (total width) NMOS devices with different finger widths, ranging from 10 to 50 μm, are extracted using this model, the results of which are shown in Figures 4.16 and 4.17. It can be seen that the model predicts substrate parameter scaling correctly.

The MOSFET model parameters (of the model in Figure 4.3) are extracted to validate the substrate model when the transistor is biased (bias condition $V_{ds} = 2.4$ V, $V_{gs} = 1.2$ V is used here). *S*-parameter simulations (see Figures 4.18–4.20) show a good agreement with the measurements for many devices ($W = 200$, 800, and 2400 μm). Additionally, Y_{22} (see Figure 4.21), which is strongly dependent on substrate parameters, exhibits a close match between simulation and measurement results.

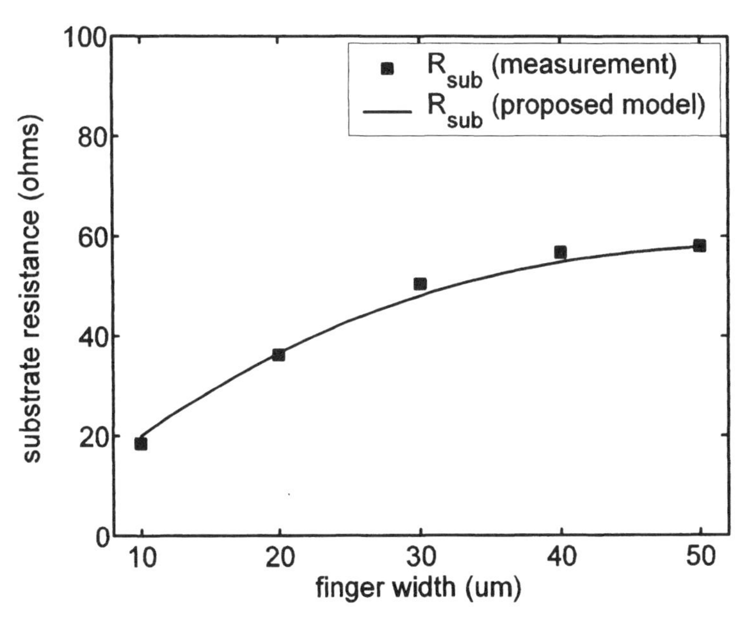

Figure 4.16. Substrate resistance of 2400-μm total width devices with different unit finger widths (V_{ds} = 0 V, V_{gs} = 0 V).

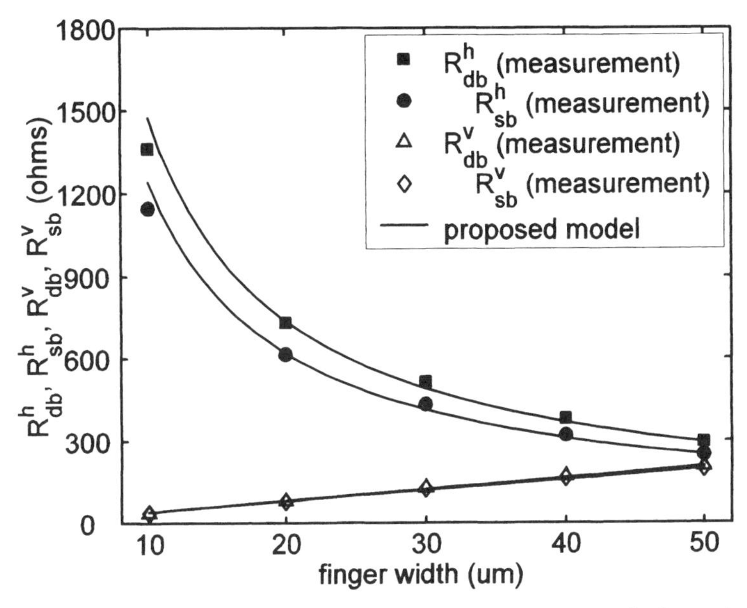

Figure 4.17. Substrate resistance components of the 2400-μm devices, in the vertical and horizontal directions (V_{ds} = 0 V, V_{gs} = 0 V).

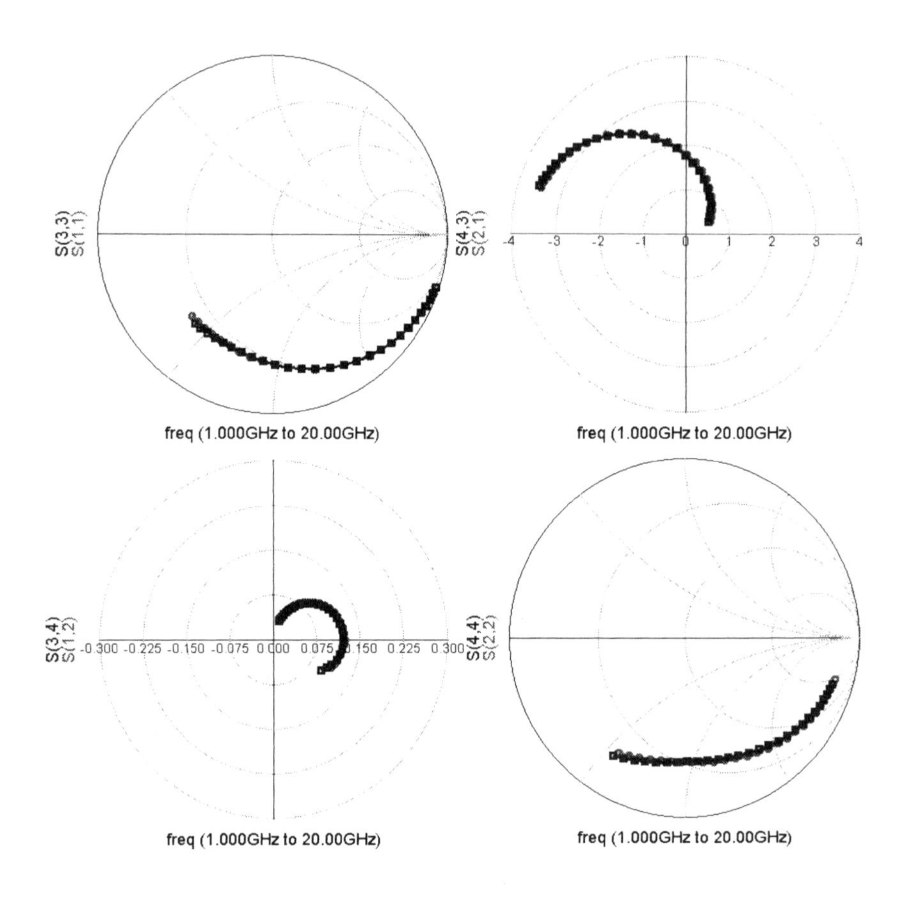

Figure 4.18. S-parameters of 0.4-μm-gate-length thick-oxide NMOS devices with width 200 μm. Circles represent measured data, and squares represent modeled data (V_{gs} = 1.2 V, V_{ds} = 2.4 V). S(1,1), S(2,1), S(1,2), and S(2,2) represent the measured S_{11}, S_{21}, S_{12}, and S_{22}, respectively, and S(3,3), S(4,3), S(3,4), and S(4,4) represent the modeled S_{11}, S_{21}, S_{12}, and S_{22}, respectively.

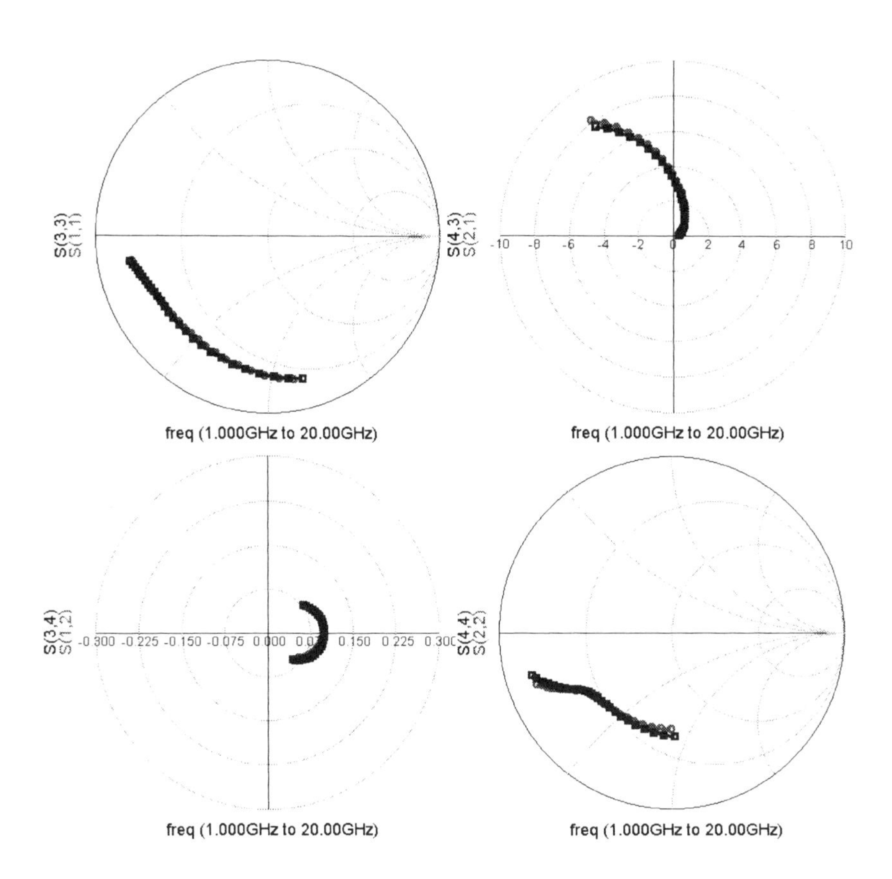

Figure 4.19. *S*-parameters of 0.4-μm-gate-length thick-oxide NMOS devices with width 800 μm. Circles represent measured data, and squares represent modeled data (V_{gs} = 1.2 V, V_{ds} = 2.4 V). *S*(1,1), *S*(2,1), *S*(1,2), and *S*(2,2) represent the measured S_{11}, S_{21}, S_{12}, and S_{22}, respectively, and *S*(3,3), *S*(4,3), *S*(3,4), and *S*(4,4) represent the modeled S_{11}, S_{21}, S_{12}, and S_{22}, respectively.

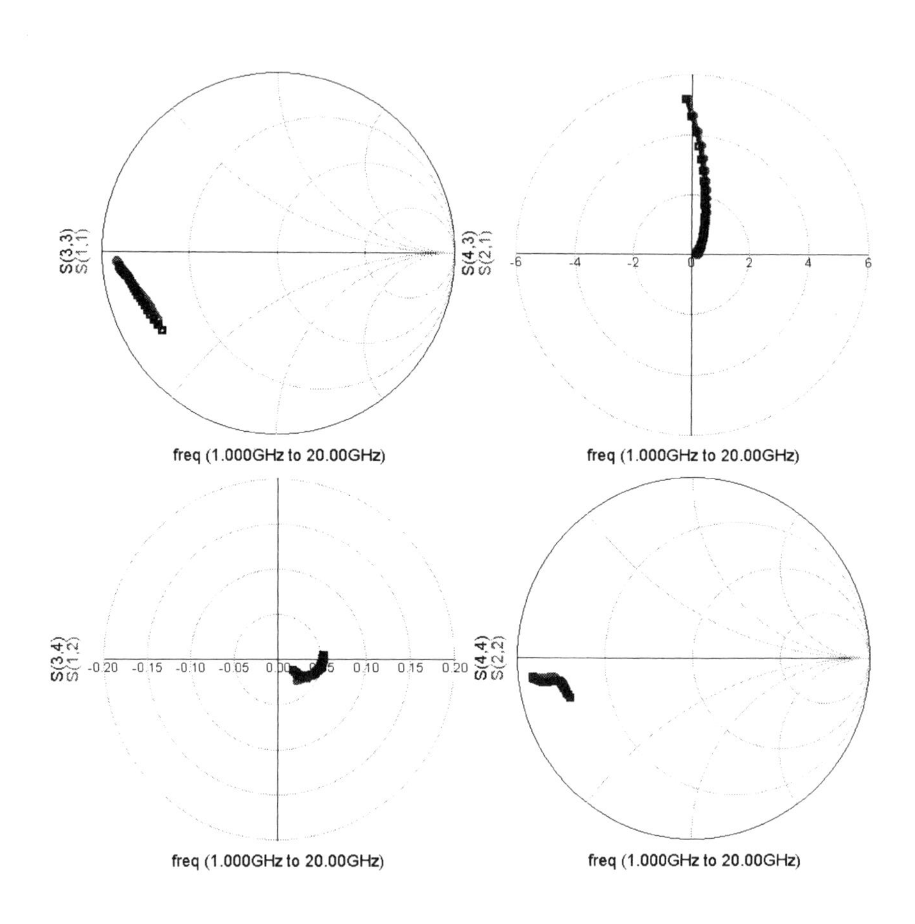

Figure 4.20. S-parameters of 0.4-μm-gate-length thick-oxide NMOS devices with width 2400 μm. Circles represent measured data and squares represent modeled data (V_{gs} = 1.2 V, V_{ds} = 2.4 V). $S(1,1)$, $S(2,1)$, $S(1,2)$, and $S(2,2)$ represent the measured S_{11}, S_{21}, S_{12}, and S_{22}, respectively, and $S(3,3)$, $S(4,3)$, $S(3,4)$, and $S(4,4)$ represent the modeled S_{11}, S_{21}, S_{12}, and S_{22}, respectively.

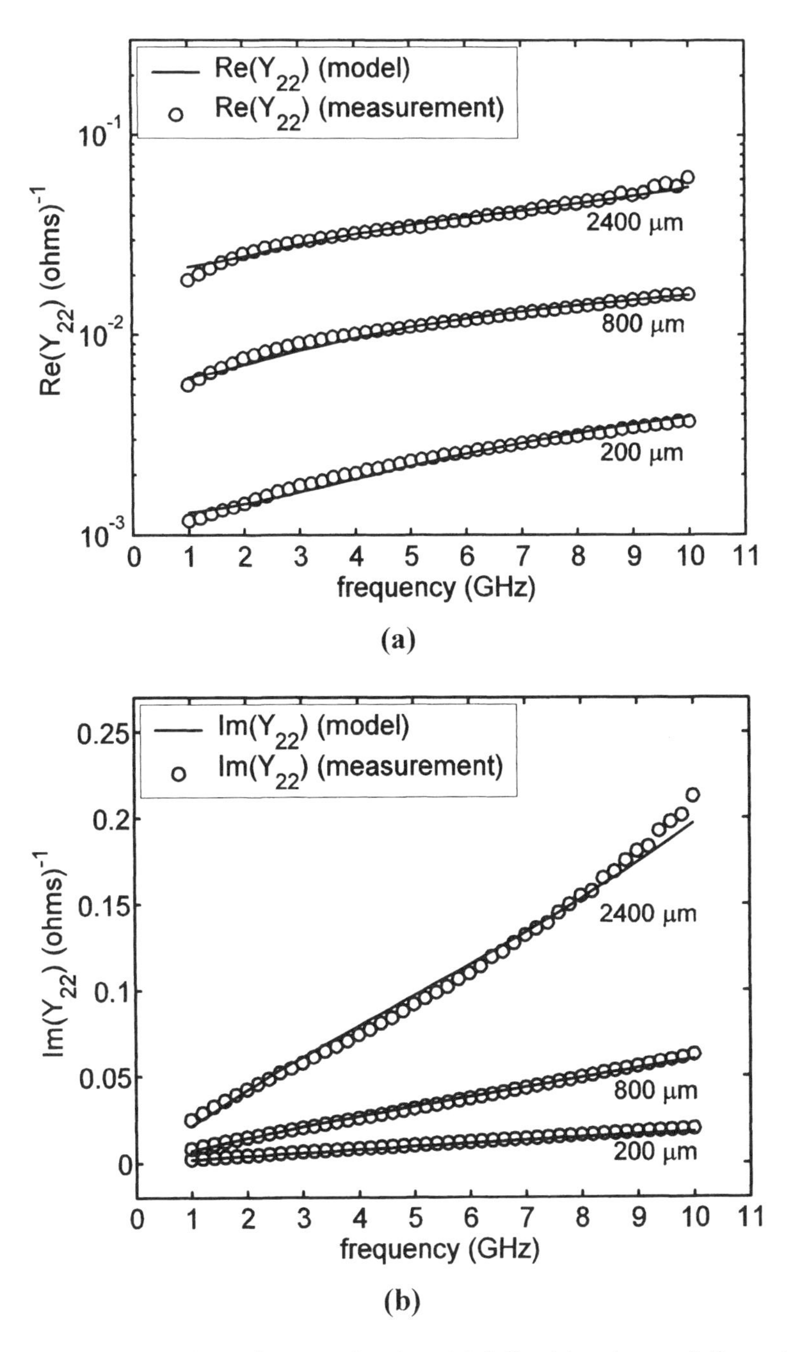

Figure 4.21. Comparison of measured and modeled Y_{22}: (a) real part of Y_{22}, and (b) imaginary part of Y_{22}.

4.2.4 Modified BSIM3v3 Model

The intrinsic parameters are readily included in the BSIM3v3 model, except for R_g and R_{sub}. While R_g can be found using the method described in the previous section, and easily added to a macromodel constructed around a regular MOSFET model, as shown in Figure 4.22, there are a few different ways of incorporating R_{sub} [7,11,21]. From the device cross section (see Figure 4.9), the substrate network can be derived analytically and modeled as shown in Figure 4.22. Here R_{db} and R_{sb} represent the combined vertical and horizontal components of R_{db} and R_{sb}. This model has junction capacitances (C_{db} and C_{sb}) that are not included in the intrinsic model (M_i). The model can be simplified to the one in Figure 4.23, which uses the BSIM3v3 model as the core. BSIM3v3 already includes intrinsic parameters and junction capacitances C_{db} and C_{sb}. The substrate resistance can be simplified to a single R_{sub} calculated from Equation 4.34. The advantage of using a single R_{sub} is the simplicity of the model. Such a model is sufficiently accurate below 10 GHz, making it suitable for most RF IC simulations.

The impedance of C_{sb} is usually higher than that of R_{sb} at <10 GHz. Therefore, when the source terminal is tied to bulk, C_{sb} can be omitted, and the equivalent model in Figure 4.3 becomes interchangeable with the modified BSIM3v3 model of Figure 4.23. The same extraction method can be applied to obtain R_{sub}. This approximation, however, may lead to deviations when device size becomes large and may vary depending on the value of W_f and L_f.

Large-signal measurements, such as, load-pull and power-sweep measurements, are used to verify the model in Figure 4.23. Load-pull measurements with the 400- and 1200-μm devices show good agreement with simulation results using this model, as seen from Figures 4.24 and 4.25. Power sweeps on the 1200- and 4800-μm devices show that the model can predict large-signal characteristics such as gain and efficiency accurately (see Figures 4.26 and 4.27). Thus, this model is suitable for a wide range of RF circuit design encompassing both small- and large-signal circuits. The scalability of the model up to large device peripheries makes it particularly attractive for use in RF power amplifier simulation and design.

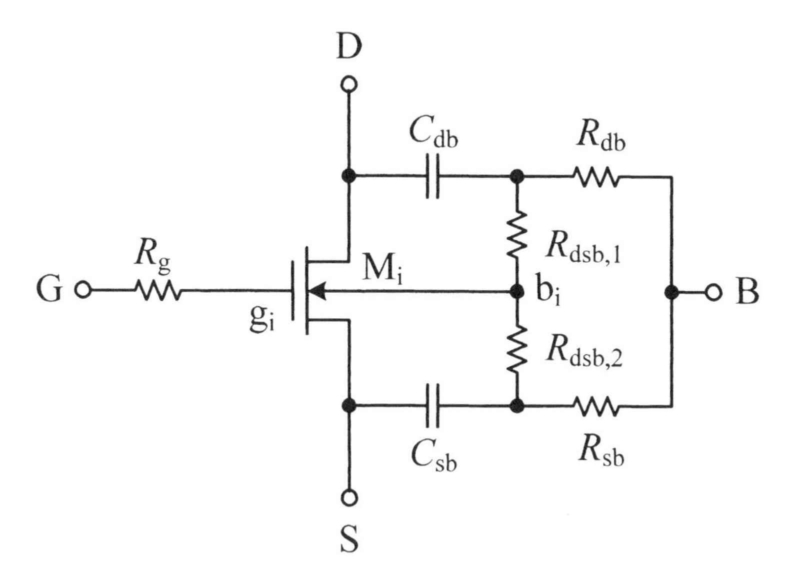

Figure 4.22. Analytical MOSFET model including gate resistance and substrate network.

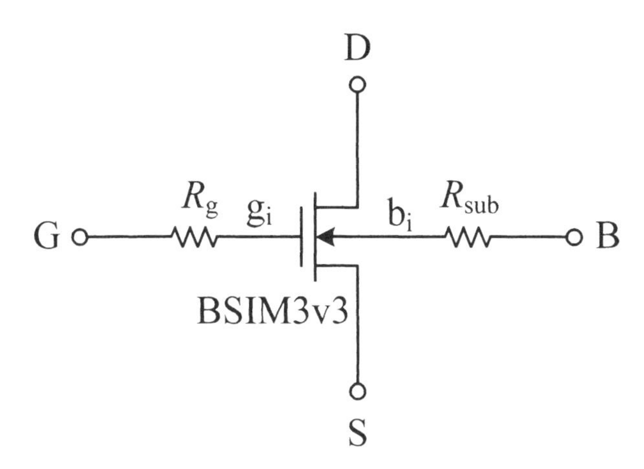

Figure 4.23. BSIM3v3 model with gate resistance and substrate network added.

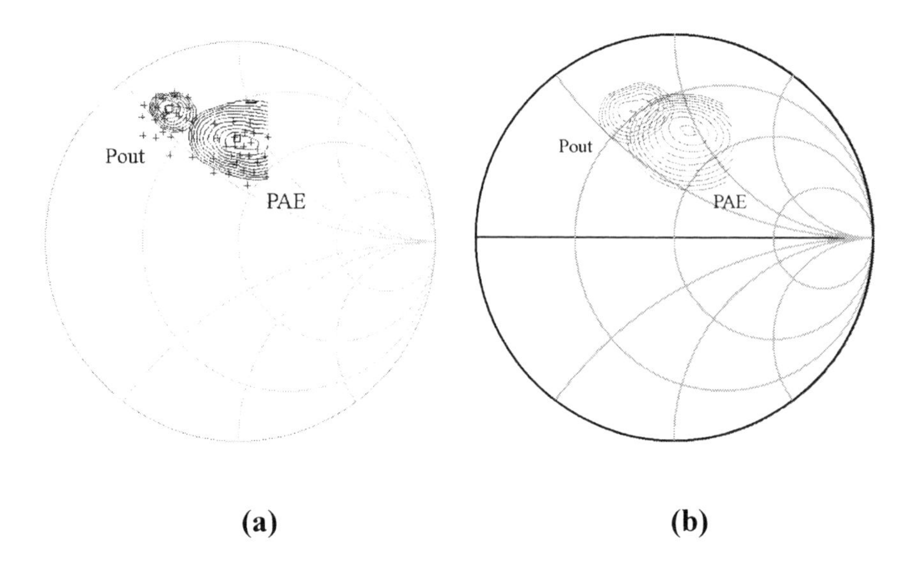

Figure 4.24. Load-pull results for the 400-μm NMOS device: (a) measurement and (b) simulation using the modified BSIM3v3 model.

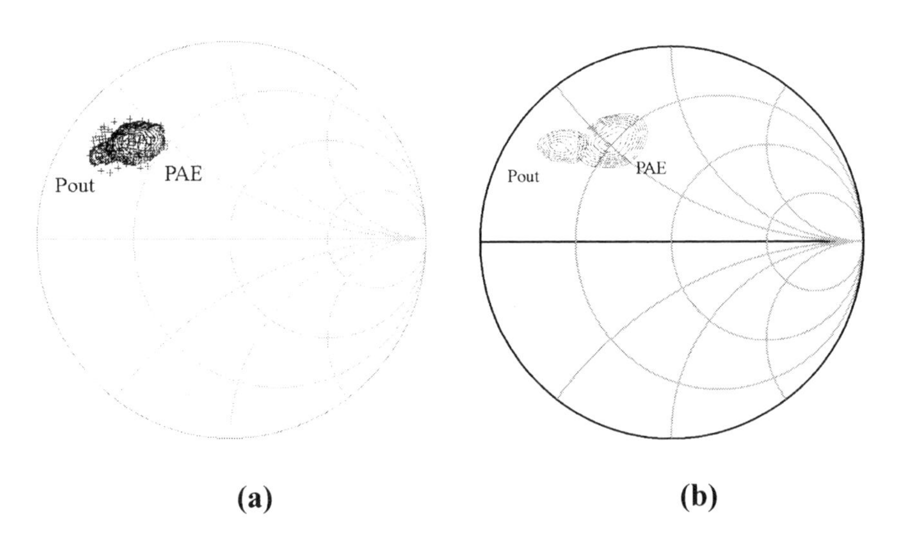

Figure 4.25. Load-pull results for the 1200-μm NMOS device: (a) measurement and (b) simulation using the modified BSIM3v3 model.

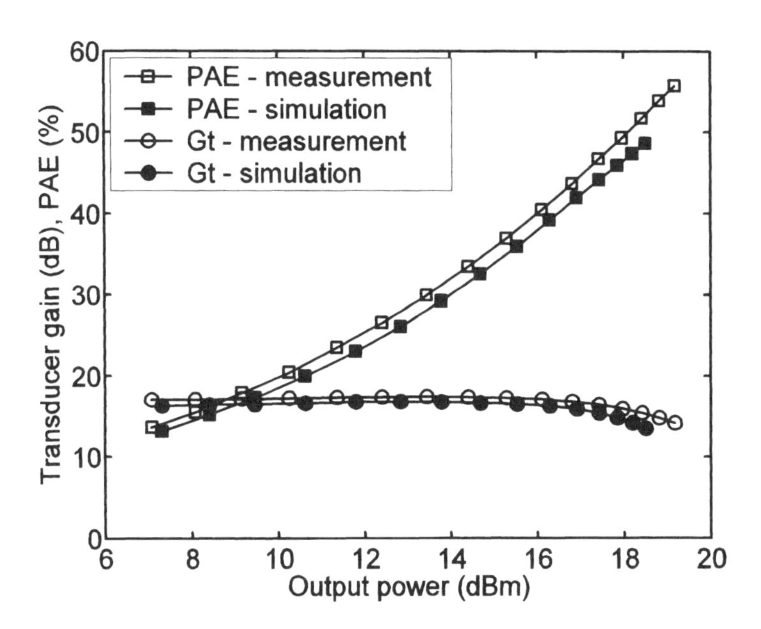

Figure 4.26. Transducer gain and power-added efficiency (PAE) of the 1200-μm device (V_{ds} = 2.4 V, V_{gs} = 0.65 V, $Z_{in,opt}$ = 5.5 + j28 Ω, $Z_{out,opt}$ = 14.5 + j23 Ω).

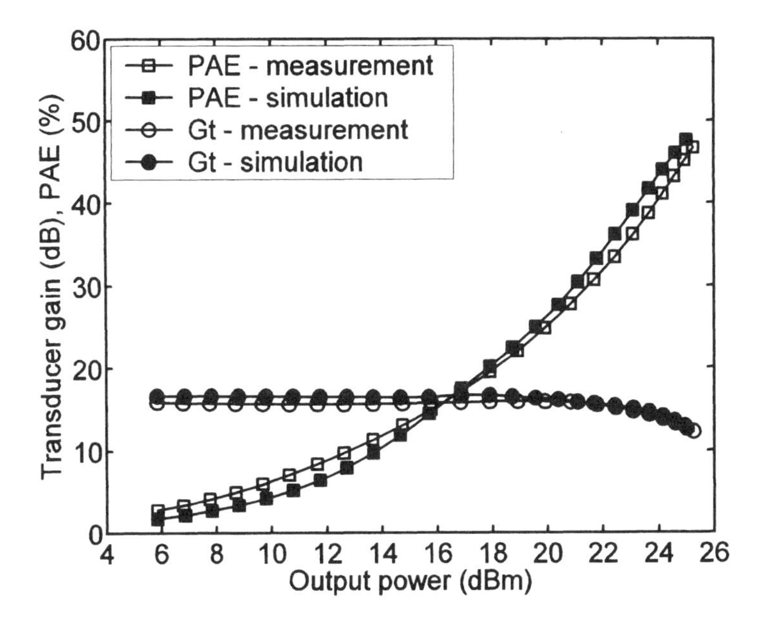

Figure 4.27. Transducer gain and power-added efficiency (PAE) of the 4800-μm device (V_{ds} = 2.4 V, V_{gs} = 0.79 V, $Z_{in,opt}$ = 2.3 + j6 Ω, $Z_{out,opt}$ = 3.9 + j4 Ω).

4.3 SUMMARY

In this chapter, considerations pertaining to substrate parameters and scaling of these parameters were discussed. Accurate substrate models, whose scalability extends to large device widths, are critical for RF power amplifier simulation. Substrate parameter scalability of RF MOSFETs having ring-shaped substrate contact surrounding the device was analyzed and modeled using analytical expressions based on device geometry. Substrate resistance and capacitance of devices up to a total gate width of 6 mm were extracted. Analysis showed that the distributed substrate coupling when using ring-shaped substrate contact can be modeled by simplifying it to substrate coupling in the vertical and horizontal directions. Using this method of modeling the substrate parameters based on device geometry, scalability of the substrate parameters can be achieved with great accuracy. The scalable substrate model was verified using small-signal *S*-parameter measurements. Further, gate resistance and the substrate network model were added to the standard BSIM3v3 model, and the simulated characteristics obtained from this modified BSIM3v3 model were found to exhibit good correlation with large-signal load-pull measurements. Thus, such a model is adequate for RF power amplifier design and simulation.

REFERENCES

1. R. Singh, A. Juge, R. Joly, and G. Morin, An investigation into the non quasi-static effects in MOS devices with on-wafer S-parameter techniques, *Proc. IEEE ICMTS*, 1993, pp. 21–25.
2. Y. P. Tsividis, *Operation and Modeling of the MOS Transistor*, McGraw-Hill, New York, 1987.
3. M. Bagheri and Y. Tsividis, A small signal dc-to-high-frequency nonquasi-static model for the four-terminal MOSFET valid in all regions of operation, *IEEE Trans. Electron Devices*, **32**: 2383–2391 (Nov. 1985).
4. K.-W. Chai and J. J. Paulos, Unified nonquasi-static modeling of the long-channel four-terminal MOSFET for large- and small-signal analyses in all operating ranges, *IEEE Trans. Electron Devices*, **36**: 2513–2520 (Nov. 1989).
5. H. J. Park, P. K. Ko, and C. Hu, A nonquasi-static MOSFET model for SPICE-transient analysis, *IEEE Trans. Electron Devices*, **36**: 561–576 (March 1989).
6. H. J. Park, P. K. Ko, and C. Hu, A charge conserving nonquasi-static (NQS) model for SPICE transient analysis, *IEEE Trans. Computer-Aided Design*, **10**: 629–642 (May 1991).
7. S. F. Tin, A. A. Osman, K. Mayaram, and C. Hu, A simple subcircuit extension of the BSIM3v3 model for CMOS RF design, *IEEE J. Solid-State Circuits*, **35**: 612–624 (April 2000).
8. C. Turchetti, P. Mancini, and G. Masetti, A CAD-oriented nonquasistatic approach for transient analysis of MOS ICs, *IEEE J. Solid-State Circuits*, **SC-21**: 827–835 (Oct. 1986).
9. P. J. V. Vandeloo and W. M. C. Sansen, Modeling of the MOS transistor for high frequency analog design, *IEEE Trans. Computer-Aided Design*, **8**: 713–723 (July 1989).

10. M. C. Ho, K. Green, R. Culbertson, J. Y. Yang, D. Ladwig, and P. Ehnis, A physical large signal Si MOSFET model for RF circuit design, *1997 IEEE MTT-S Int. Microwave Symp. Digest*, June 1997, pp. 391–394.
11. J.-J. Ou, X. Jin, I. Ma, C. Hu, and P. Gray, CMOS RF modeling for GHz communication ICs, *VLSI Technology Symp. Digest*, June 1998, pp. 94–95.
12. B. Razavi, R. H. Yan, and K. F. Lee, Impact of distributed gate resistance on the performance of MOS devices, *IEEE Trans. Circuits Syst. I*, **41**: 750–754 (Nov. 1994).
13. Technology Modeling Associates, *Medici*, 1997.
14. Y. Cheng, and M. Matloubian, On the high-frequency characteristics of substrate resistance in RF MOSFETs, *IEEE Electron Device Lett.*, **21**: 604–606 (Dec. 2000).
15. S. Lee, A small-signal RF model and its parameter extraction for substrate effects in RF MOSFETs, *IEEE Trans. Electron Devices*, **48**: 1374–1379 (July 2001).
16. S. Lee, Direct extraction of substrate parameters for the small-signal model of a RF MOSFET, *Proc. Asia-Pacific Microwave Conf.*, Nov. 2003, vol. 1, pp. 346–349.
17. Y. Cheng and M. Matloubian, Parameter extraction of accurate and scalable substrate resistance components in RF MOSFETs, *IEEE Electron Device Lett.*, **23**: 221–223 (April 2002).
18. C. C. Enz and Y. Cheng, MOS transistor modeling for RF IC design, *IEEE J. Solid-State Circuits*, vol. 35, pp. 186-201, February 2000.
19. J. Han, and H. Shin, A scalable model for the substrate resistance in multi-finger RF MOSFETs, *IEEE MTT-S Int. Microwave Symp. Digest*, 2003, vol. 3, pp. 2105–2108.
20. S. H.-M. Jen, C. C. Enz, D. R. Pehlke, M. Schroter, and B. J. Sheu, Accurate modeling and parameter extraction for MOS transistors valid up to 10 GHz, *IEEE Trans. Electron Devices*, **46**: 2217–2227 (Nov. 1999).
21. W. Liu, RF MOSFET modeling accounting for distributed substrate and channel resistances with emphasis on the BSIM3v3 SPICE model, *IEDM Tech. Digest*, 1997, pp. 309–312.
22. D. Pehlke, High frequency application of MOS compact models and their development for scalable RF model libraries, *Proc. IEEE Custom Integrated Circuit Conf.*, May 1998, pp. 219–222.
23. R. T. Chang, M.-T. Yang, P. P. C. Ho, Y.-J. Wang, Y.-T. Chia, B.-K. Liew, C. P. Yue, S. S. and Wong, Modeling and optimization of substrate resistance for RF-CMOS, *IEEE Trans. Electron Devices*, **51**: 421–426 (March 2004).
24. N. Srirattana, D. Heo, H.-M. Park, A. Raghavan, P. E. Allen, and J. Laskar, A new analytical scalable substrate network model for RF MOSFETs, *IEEE MTT-S Int. Microwave Symp. Digest*, 2004, pp. 699–702, 2004.
25. N. Srirattana, D. Heo, A. Raghavan, K. Lim, P. E. Allen, and J. Laskar, A scalable small-signal model for MOSFET including substrate network, *Proc. Asia-Pacific Microwave Conf.*, Nov. 2003, vol. 2, pp. 1203–1206.
26. D. M. Pozar, *Microwave Engineering*, Wiley, New York, 1999.

5

POWER AMPLIFIER IC DESIGN

5.1 INTRODUCTION

The RF power amplifier amplifies a modulated RF signal for transmission by an antenna, consuming DC power in the process. The primary aim of the RF power amplifier is to transmit the required signal with sufficient power to be sensed by the receiver. The power level is determined by the intended application, as well as the location of the power amplifier in the overall system. For instance, power amplifiers in cellular basestations should be able to output tens to hundreds of watts. For satellite communications, this can be of the order of thousands of watts. For portable wireless handsets and wireless LAN transmitters, the power requirement is significantly lower, usually of the order of a few hundred milliwatts. The information being transmitted is encoded from the baseband signal to an RF bandpass signal, prior to transmission. The spectrum of the transmitted signal has to fit within a spectrum mask imposed by the Federal Communications Commission (FCC). The purpose of the mask is to prevent interference with signals in adjacent channels. The transmitted bandpass signal is modulated using one of several modulation schemes. The particular modulation scheme used for a given application affects the linearity requirement of the power amplifier. In general, modulation schemes can be classified into two categories: constant-envelope and nonconstant-envelope schemes. In constant-envelope modulation schemes, which are used in applications like the global system for mobile communications (GSM), and digital enhanced cordless telecommunications (DECT), there is no information contained in the amplitude of the transmitted signal. Hence a nonlinear power amplifier can be used in this case, the direct benefit being better power amplifier operating efficiency. Nonconstant-envelope modulation schemes, such as those used in code-division multiple access (CDMA), and wideband CDMA (WCDMA) systems, contain information in the amplitude of the transmitted signal in the form of amplitude modulation. While this enhances the spectral efficiency of the signal, with more information transmitted for the same frequency bandwidth, it has the unfortunate effect of placing a stringent demand on the linearity of the power amplifier. This is usually achieved at the expense of the power-added efficiency (PAE) of the power amplifier.

5.2 POWER AMPLIFIER DESIGN METHODOLOGY

A power amplifier could be constructed based largely on data obtained from measurements on test structures. In such an approach, power-cell test structures of various dimensions and configurations, are designed, and the sizing of the amplifier stages, and terminations for the best output power, efficiency, and linearity, are determined by source- and load-pull measurements. The optimum terminations that meet the desired specifications are decided according to a compromise among the power, efficiency, and linearity match conditions; and input, output, and interstage matching networks are designed to achieve these terminations. Test structures may also be used to determine the optimum unit cell configuration, distance between transistor fingers, spacing between unit cells, and the best arrangement of transistors in the power cell, from a thermal perspective. While such an experimental approach is quite useful, it is very time- and cost-intensive. Also, since the optimum match impedances of large power transistors are typically very small, in practice, the accuracy of load-pull techniques is limited in the absence of sophisticated equipment such as prematching and harmonic tuners, which add to the development cost. Therefore, power amplifiers are usually initially designed using CAD simulators, if reasonably accurate models are available (it is worth noting here that typical transistor models such as the SPICE Gummel–Poon model are usually inadequate to predict power amplifier performance with sufficient accuracy, since they do not model effects such as self-heating, which are important from a power amplifier perspective), or even using fundamental load-line principles, and the performance of the resulting amplifier is optimized using load-pull techniques (This usually involves tuning the output match impedance to optimize performance.) A reasonable compromise is the use of a combination of simulations and fabricated test structures to arrive at an optimum design with a minimum number of iterations. Test structures would be particularly useful in testing layout strategies for the power stages, especially from a thermal perspective, since these effects do not scale well with typical transistor models. As more complete and scalable transistor models become available, the need for post-fabrication experimental tuning is expected to decrease. In this context, it is worth noting that a scalable transistor model for power amplifier design needs to be devised with a particular layout topology for the power cell in mind. This is because the scaling of the transistor device itself does not include the scaling of the interconnect parasitics associated with the power cell. Also, a scalable thermal model is required, and this is expected to be largely empirical.

The design of matching networks is a critical aspect of RF power amplifier design. The output match network is usually off-chip, and is realized using a combination of surface-mount components and transmission line structures drawn on the printed circuit board (PCB). Sometimes, passives embedded in the board or within the package substrate are used, especially if it is a high-quality material. While the optimum output match impedance for maximum power typically deviates from the conjugate match impedance, the interstage match between the output stage and the preceding driver stage is also likely to do so, since in many cases, the output of the driver stage is large-signal. The design of interstage matching networks is equally important in realizing the best possible performance from the power amplifier, and this is often not very easy since these matching networks are mostly on-chip, thereby reducing the type, range of values, and quality of components available. In addition to these considerations, estimating and including the effect of bond wires, bond pads, and package parasitics during the design phase is important in RF power amplifier design. Apart from the design of the power amplifier IC itself, the design of the package and the board are equally important. Careful attention must be paid to thermal management, which can otherwise degrade the performance of the power amplifier significantly. Chapters 6 and 7 include several practical power amplifier design examples, which, in addition to introducing advanced techniques, also illustrate the various fundamental considerations in the design of an RF power amplifier.

5.3 CLASSES OF OPERATION

Power amplifiers are classified into classes according to their mode of operation. There are several recognized lettered classes (A, B, C, etc.), with quite a few combinations of these classes adding to the variety (and confusion!). If the AC current amplitude never exceeds the quiescent DC value, the power amplifier is said to be operating in Class A. Otherwise, depending on the conduction angle (θ) of a sinusoid at the collector node, which is the portion of the cycle for which the current is greater than zero, the power amplifier is said to be in Class AB ($\pi < \theta < 2\pi$), Class B ($\theta = \pi$), or Class C ($0 < \theta < \pi$). The current and voltage waveforms corresponding to these classes of operation are shown in Figure 5.1. Note that the classes of operation are based on the conduction angle of the current waveform. Class A provides linear amplification of the input signal, since there is no clipping or distortion of the signal. However, the maximum possible drain efficiency is 50%. In practice, because of various nonidealities and parasitic effects, it is possible to realize an efficiency of only about 30%, at most. Further, at low output power levels, that is, when the swing of voltage and current waveforms is much less than their quiescent values, this efficiency drops further. This is of significance and concern in many practical applications (for example, CDMA) where the power amplifier transmits at peak output power for only a small portion of the time.

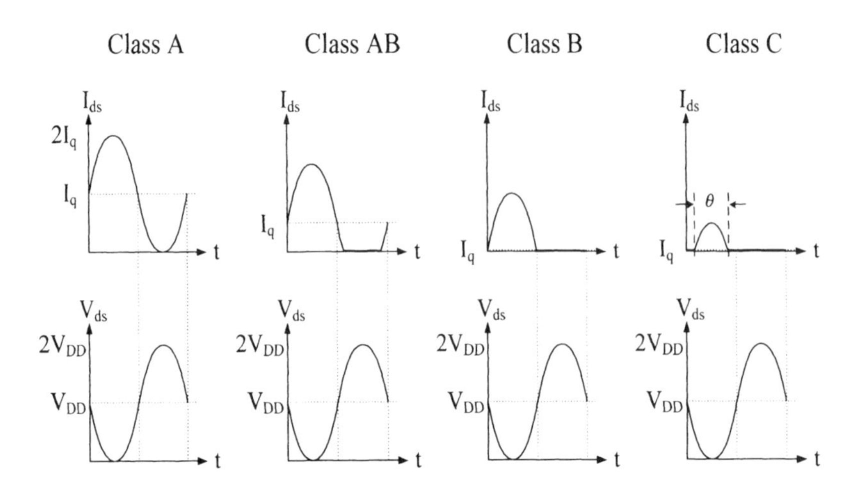

Figure 5.1. Voltage and current waveforms for different classes of operation of a power amplifier.

In Class B operation, the amplifier draws current only during one-half of the input drive interval and thus, the transistor is on for half the cycle. Therefore, the transistor consumes less power than in Class A operation and is more efficient. A tuned Class B amplifier or two Class B amplifiers in a push–pull configuration can be used for linear amplification. However, in practice, the distortion that occurs when the output "crosses over" from one amplifier to the other in a push-pull configuration, limits the linearity of this topology. Because of limitations in the practical realization of Class B amplifiers, and the poor efficiency of Class A amplifiers, most linear RF power amplifiers, especially the ones in mobile handsets, are operated in moderate Class AB. Since the actual power output level in a typical nonconstant-envelope modulation system (CDMA, for instance), is much smaller than the peak level most of the time, sufficient linearity is still obtained if the amplifier is biased in Class AB with respect to the peak output level.

It was mentioned before that a Class B amplifier could be used as a linear amplifier. This is because an ideal Class B amplifier is actually linear from an amplitude-modulated signal perspective, since a reduction in input drive power results in an equivalent reduction in the output power, as the conduction angle remains constant. However, the efficiency of the amplifier also decreases with decreasing output power. An ideal Class AB amplifier, on the other hand, is nonlinear, while a Class A amplifier is linear. It is important to note that these conclusions are valid only for ideal amplifiers. In reality, the linearity of a particular mode of operation is affected by the nonlinearity of the transconductance characteristic of the transistor device itself. If the transfer characteristic of the transistor is nonlinear, then a Class A or a Class B amplifier may become nonlinear for an amplitude-modulated signal. In reality, it turns out that amplifiers biased in Class AB often provide the best linearity for amplitude modulated signals, and since they are advantageous from the efficiency point of view as well, Class AB power amplifiers are often used in mobile applications using amplitude modulation.

The maximum drain efficiency (η) of an amplifier can be calculated as a function of the conduction angle from the following expression [1]:

$$\eta = \frac{1}{4}\left[\frac{\theta - \sin(\theta)}{\sin\left(\frac{\theta}{2}\right) - \frac{\theta}{2}\cos\left(\frac{\theta}{2}\right)}\right] \tag{5.1}$$

The maximum drain efficiency of a Class A amplifier ($\theta = 360°$) is 50%, and the efficiency increases for smaller conduction angles, through Class AB, with Class B ($\theta = 180°$) providing a drain efficiency of 78.5%, while Class C amplifiers can ideally reach an efficiency of 100% (when $\theta = 0°$). However, the output power to the load will drop rapidly when θ approaches zero, as seen from the expression

$$P_{out} \propto \frac{\theta - \sin(\theta)}{1 - \cos\left(\frac{\theta}{2}\right)} \tag{5.2}$$

Therefore, it is not possible to use Class C bias to achieve 100% efficiency while providing full output power. Class C amplifiers are impractical for solid-state power amplifiers, because the large negative swing of the input voltage, which coincides with the collector output voltage peak, results in the worst condition for reverse breakdown in any kind of transistor, and even small leakage currents flowing at this point of the cycle have a detrimental effect on the efficiency. Class C, however, was very much the preferred mode in the days of vacuum-tube amplifiers.

For constant-envelope operation, efficiency can be improved by the use of harmonics to shape the current or voltage waveform. Such an approach can be applied to Class A operation by overdriving the input signal, thereby shaping the drain voltage toward a square wave. This so-called overdriven Class A amplifier is different from Class AB since the transistor is still biased in Class A (bias point is located midway between cutoff and saturation). In this case, the drain voltage has sharper edges (smaller rise and fall times) than does a normal sinusoidal waveform, thus reducing the overlap between output current and voltage waveforms, and eventually increasing the efficiency. Class AB and Class B amplifiers can also be overdriven, as can Class C amplifiers. However, output power and efficiency gains are typically observed only in the case of Class A and low Class AB (i.e., conduction angle close to Class A operation).

Class B and Class C amplifiers tend to show a decrease in efficiency with increasing overdrive.

The concept of reducing the overlap between voltage and current waveforms has been applied to operate the transistor as a switch, sometimes called a "switching amplifier." In an ideal switch, current and voltage are mutually exclusive in time, therefore contributing zero power dissipation. This concept is used in Class D and E amplifiers. Class D amplifiers use transistors to switch between "A" and "B" in Figure 5.2a, resulting in a square waveform across the switch without coexisting current. Each switch conducts a half-cycle of the current, and these are combined to form the full sine wave current I_0. Class D is normally used in the low-frequency range (10–30 MHz), where the parasitic reactance of the switch is not an issue. Practical problems include control of the switches, which must be synchronous with no overlapping interval, to avoid leakage current and unwanted DC dissipation. Also, even though the ideal efficiency of Class D amplifier is 100%, the nonzero saturation voltage of the transistor causes power dissipation and reduces efficiency. A push–pull implementation of a switching Class D amplifier is shown in Figure 5.2b, and ideal current and voltage waveforms of this class of amplifier are shown in Figure 5.2c.

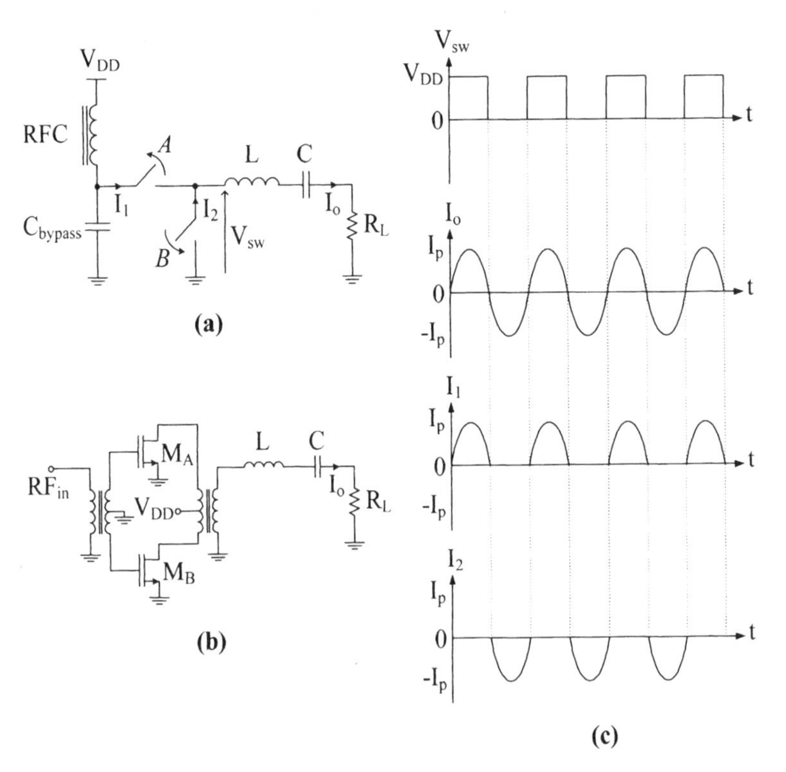

Figure 5.2. Class D operation: (a) schematic diagram; (b) push–pull Class D amplifier; and (c) voltage and current waveforms in Class D operation.

The Class E amplifier (Figure 5.3) uses a transistor to switch the current flowing through itself and the current charging the output capacitor (C_p) to create output voltage. The voltage waveform across C_p is given by (see also Figure 5.3 b),

$$v_C(\gamma) = \left(\frac{1}{\omega C_p}\right) \int (2\pi - \beta)^{\gamma} i_C(\gamma) d\gamma \tag{5.3}$$

The voltage peaks at the zero crossing of the capacitor current and is a function of the selected angle (α) and closing time of the switch (β). The performance of the Class E amplifier is dependent on these factors and is also strongly determined by the design of the output match. L_s and C_s form a resonant circuit to filter the sine wave output to the load R_L. C_p is important in determining the peak voltage and the maximum frequency for the circuit to operate with 100% efficiency. It has to satisfy the condition that no voltage and current overlap, which means that C_p must be fully discharged before the next switching cycle starts. A significant downside of the Class E amplifier is that the peak voltage (V_p) can be as high as 3–5 times the supply voltage. This requires the use of high-breakdown transistors or other special techniques, such as the stacking of transistors. Detailed analysis of the efficiency of Class E amplifiers can be found in the literature [2,3]. Class E operation can achieve 100% efficiency in theory, but is difficult to realize with solid-state power amplifiers at radiofrequencies, because the parasitics associated with a solid-state transistor render it a rather poor approximation to an ideal switch.

Another class of switching amplifier, namely, Class F, has similarities to the overdriven Class A amplifier. In the typical overdriven Class A amplifier, clipping of both the voltage and current output waveforms is allowed, whereas in Class AB and Class B overdriven amplifiers, the output harmonics are shorted, resulting in a sinusoidal output voltage. In Class F, the output voltage waveform is allowed to be shaped, as in overdriven Class A. This stems from the observation that if the output voltage of an amplifier biased in Class B is shaped toward a square wave, significant improvement in efficiency is achieved. This "squaring" of the output voltage waveform of the amplifier is accomplished by presenting appropriate harmonic terminations at the output of the power amplifier. The ideal case, where the output voltage waveform is a square wave, containing all the odd harmonics, but no even harmonic components, is the previously described Class D mode of operation. Thus, the output of the Class D amplifier is presented with a short at all even harmonics, and an open at all the odd ones. The current waveform is ideally half-sinusoidal, and contains only even harmonic components. It should be noted that while the Class D amplifier described here has output waveforms similar in shape to the switching Class D amplifier described previously, the topology described here is different in that Class D operation is achieved here using appropriate harmonic terminations. The distinction is rendered somewhat irrelevant by the fact that this case is clearly not feasible at microwave frequencies, because it is practically impossible to realize the required harmonic terminations for the higher harmonics. However, a configuration with reduced harmonic termination requirements shows promise for RF operation. This is the so-called Class F mode (Figure 5.4), in most practical applications of which the second harmonic termination is a short, and the third is an open. More generally, an amplifier with output voltage having the fundamental and any finite number of odd harmonics, and output current having the fundamental and any finite number of even harmonics, is said to operate in Class F mode. Practically, a quarter-wave transmission line is used to short even harmonics in addition to the supplying bias current.

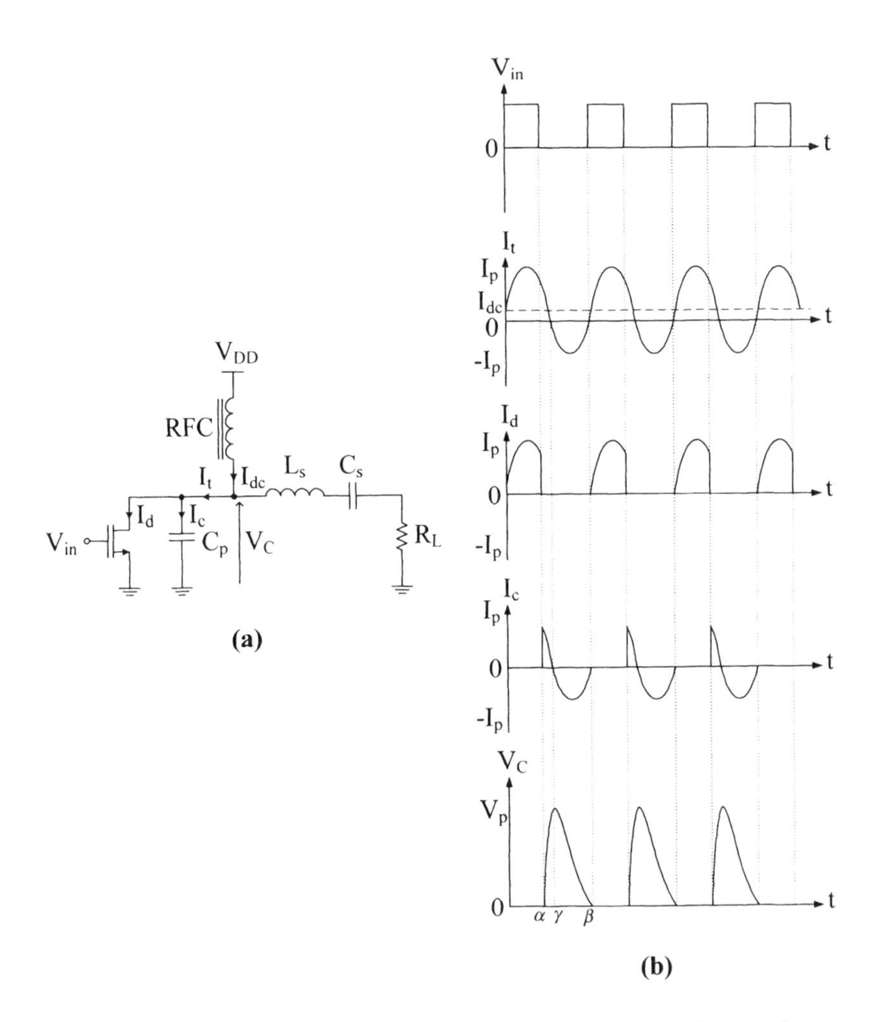

Figure 5.3. Class E operation: (a) schematic diagram; and (b) voltage and current waveforms of a Class E amplifier.

An advantage of the Class F amplifier over the Class E amplifier is that the maximum drain voltage is just twice the supply voltage in a Class F amplifier, unlike typically 3–5 times in Class E. Yet another class of operation is the so-called inverse Class F, where the harmonic terminations are reversed with respect to Class F, that is, the third-harmonic termination is a short, while the second-harmonic one is an open, in practical implementations. The terms *odd* and *even Class F* are sometimes used, instead of *Class F* and *inverse Class F*, respectively (The *odd* and *even* refer to the harmonic content of the output voltage waveform.) In both Class F and inverse Class F, the transistor is biased at cutoff, as in Class B. The popular Class F topology with second and third harmonic resonators, theoretically has an efficiency of 75%, with maximally flat waveforms [4]. Maximally flat waveforms are a good approximation to the waveforms in a real Class F power amplifier, and simplify mathematical analysis. The term *maximally flat* refers to the fact that the derivatives of these waveforms are zero at their peak values. From such an analysis, it is found that including harmonics through the fifth increases the efficiency to about 83%.

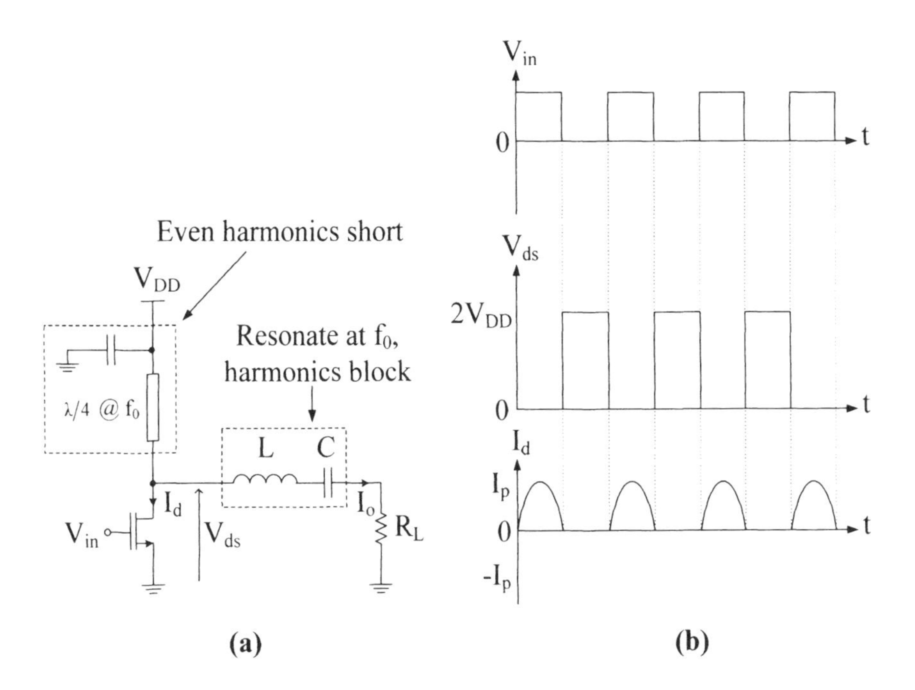

Figure 5.4. Class F operation: (a) schematic diagram; and (b) voltage and current waveforms in Class F operation.

5.4 PERFORMANCE METRICS

The performance of RF power amplifiers is characterized by a unique set of performance metrics. Therefore, it is useful for the RF power amplifier designer to understand the definition and significance of these figures of merit. The power levels in RF power amplifier design are generally described in dBm, which is the decibel representation of the power referenced to 1 milliwatt (mW). The efficiency measure used in the previous section on power amplifier classes is more specifically called output or drain efficiency, and is defined as the ratio of output AC power and DC power consumption. In RF power amplifiers, the input drive is also usually substantial. In order to take this into account, the metric, power-added efficiency (PAE) is used, and is defined as,

$$PAE = \frac{P_{\text{out}} - P_{\text{in}}}{P_{\text{dc}}} \tag{5.4}$$

where P_{out} and P_{in} are the output and input RF power levels, respectively, and P_{dc} is the DC power consumption. An important consideration that is often overlooked when quantifying the efficiency of a power amplifier is the effect of loss in various support circuits, like harmonic filters, or resistive losses associated with couplers, FETs, or sense resistors used in power control loops. At high output power levels, a loss of even a couple of tenths of 1 dB can lower the effective PAE of the power amplifier at the antenna by 10–15%.

Various metrics are of interest, when considering the linearity of a power amplifier. Two of these are the 1 dB compression point, and the third-order intermodulation (IM3) distortion. The gain of a power amplifier is another important performance metric, and linearity of a power amplifier, from an amplitude-modulated signal perspective, can be characterized by the constancy of its gain with input drive level. The 1-dB compression point is defined with respect to a sinusoidal (or single tone) input. In general, the gain of a nonlinear system is dependent on the amplitude of the sinusoidal input, and decreases with increasing input amplitude. This can be analyzed by considering the following familiar Taylor series expansion for the output of a memoryless nonlinear device with respect to its input:

$$V_{\text{o}} = a_0 + a_1 V_{\text{i}} + a_2 V_{\text{i}}^2 + a_3 V_{\text{i}}^3 + \cdots \tag{5.5}$$

Now, if the input V_i is a single tone, i.e. $V_{\text{i}} = A\cos(\omega t + \phi)$, then, from Equation 5.5, the fundamental component of the output is

$$V_{\text{o,fund}} = a_1 A \cos\omega t + 0.75 a_3 A^3 \cos\omega t + \cdots \tag{5.6}$$

Typically a_3 is negative, so as the amplitude A increases, the gain decreases. The 1-dB compression point is the point at which the gain of the power amplifier is 1 dB less than its small-signal gain. In physical devices, as the input level increases, the output clips and saturates, that is, it reaches the maximum that can be provided by the device. The variation in gain with input amplitude, also referred to as *AM-AM conversion*, degrades amplitude-modulated signals and increases adjacent channel leakage. It is important to note that, in addition to gain compression, AM-AM distortion is also caused by gain expansion, which is an increase in gain, sometimes observed in power amplifiers, below the saturation level. Power amplifiers also cause phase distortion at high amplitude, an effect called *AM-PM distortion*. This also results in signal degradation and adjacent-channel leakage, although the latter is usually dominated by AM-AM distortion. Linear power amplifiers aim to operate at power levels lower than the 1-dB point, to achieve high linearity. In other words, the power amplifier design needs to be such that the 1-dB point is comparable to, or greater than, the maximum required power level. From Equation 5.5, for a sinusoidal input, it is evident that the output contains components at the harmonics of the fundamental frequency, that is, it produces harmonic distortion. Harmonics can be suppressed using filters.

Intermodulation distortion is defined with reference to two (or more) input tones. If two tones (say, at frequencies f_1 and f_2) are input to a nonlinear system, in general, it produces additional tones or distortion products at the output. Of these, the so-called third-order intermodulation (IM3) products, located at $2f_1$-f_2 and $2f_2$-f_1, are the most important. The fifth-order intermodulation (IM5) products are located at $3f_1$-$2f_2$ and $3f_2$-$2f_1$, and so on. The IM3 products are of particular interest because, for two closely spaced input tones, the IM3 products are located very close to the original input frequencies. The slope of the IM3 product versus input power curve is 3 times that of the fundamental–input power curve, and the theoretical point at which the two extrapolated curves intersect (in reality the curves saturate), is called the *third-order intercept*, and is often used as a figure of merit for linearity.

While intermodulation products are useful indicators of power amplifier linearity, two sinusoidal tones do not adequately represent actual transmitted RF signals. The spectrum (or power spectral density) of a real-world input signal to an RF power amplifier is bandpass in nature, and is the result of the modulation of the amplitude, phase or frequency of the RF carrier using one of several modulation techniques. Thus, other metrics have been developed to characterize the linearity of a power amplifier in real applications. One such figure of merit is the *adjacent-channel power ratio* (ACPR), also called *adjacent-channel leakage ratio* (ACLR). ACPR is the ratio between the power in the adjacent channel and that in the main channel (see Figure 5.5), and is measure of the degradation in spectral regrowth caused by the power amplifier.

To understand spectral regrowth intuitively, consider the spectrum of a digitally modulated signal, a typical representation of which is shown in Figure 5.5. The main channel of the signal, with Nyquist bandwidth f_{BW}, is between f_1 and f_2. Intuitively, the signal may be considered to consist of a number (or a continuum) of closely spaced tones between f_1 and f_2. Note that such a description of the signal is mathematically inaccurate, and is used only to develop an intuitive understanding of spectral regrowth. Let us now consider the IM3 products resulting from these several closely spaced input tones. The extent of spread of the IM3 products in frequency is given by the intermodulation products generated by the tones at the main channel edges (i.e., f_1 and f_2), and these IM3 products are located at f_1-f_{BW} and f_2+f_{BW}. Thus, the IM3 products spread out one Nyquist bandwidth (f_{BW}) from each edge of the main channel. The adjacent channel is f_{BW} wide, as shown in Figure 5.5. Similarly the IM5 products spread out $2f_{BW}$ from each main-channel edge. The channel two bandwidths away from the center of the main channel, that is, at an offset of two bandwidths from the carrier frequency, is sometimes called the *alternate channel* (see Figure 5.5). The leakage ratio for this channel with respect to the main channel is referred to as ACPR2 or ACLR2 (the ACPR for the adjacent channel is then called ACPR1 to distinguish it from the alternate channel).

Spectral regrowth, which is caused by nonlinearity in the RF transmitter circuits as described above, is a major concern in wireless transmission. To limit interference between signals in adjacent frequency bands and the consequent signal degradation, there are specified limits to spectral regrowth, given by the ACPR specification. A "spectrum mask" is usually specified for a wireless communication standard, and the spectral regrowth skirts must fit within the bounds defined by this mask. ACPR varies with modulation format, and is also affected by other factors; for instance, it depends on how the channel is loaded [5]. This dependence is due largely to the statistical distribution of power level with respect to time. For example, in CDMA systems, the peak-to-average power ratio (PAPR) statistics is dependent on various factors, such as number of code channels, and the code channel number assignments. Thus, the stimulus signal needs to be well defined for consistent and repeatable measurements, and must adequately represent real-world signals. Complementary cumulative distribution function (CCDF) curves [6] are used to specify the power statistics of the signal, for this purpose. A CCDF curve is a plot of relative power level versus probability, and shows the percent of time the signal spends at or above any power level within the dynamic range of the signal. Multicarrier signals, used in some wireless communication systems, have more demanding CCDFs than do single-carrier signals, and require more linearity from the power amplifier. Such signals exhibit very high peak-to-average power ratios. An appropriate stimulus must be used to correctly characterize the ACPR of power amplifiers used in such systems.

Noise power ratio (NPR) is another linearity metric, which is particularly useful in satellite systems, where different modulation formats might be present. As the number of carriers increases, the CCDF of the signal approaches that of additive white Gaussian noise (AWGN). The NPR test uses a Gaussian noise stimulus with a notch located in the center of the band to remove a slice of the input signal spectrum. The stimulus is generated using an arbitrary waveform signal generator capable of generating a multitone signal, with tones omitted at the frequencies where the notch is desired. The ratio of the power spectral density of the signal (plus the noise, which is negligible) and that of the noise in the notch, at the output, is the noise power ratio. It is a measure of in-band intermodulation distortion.

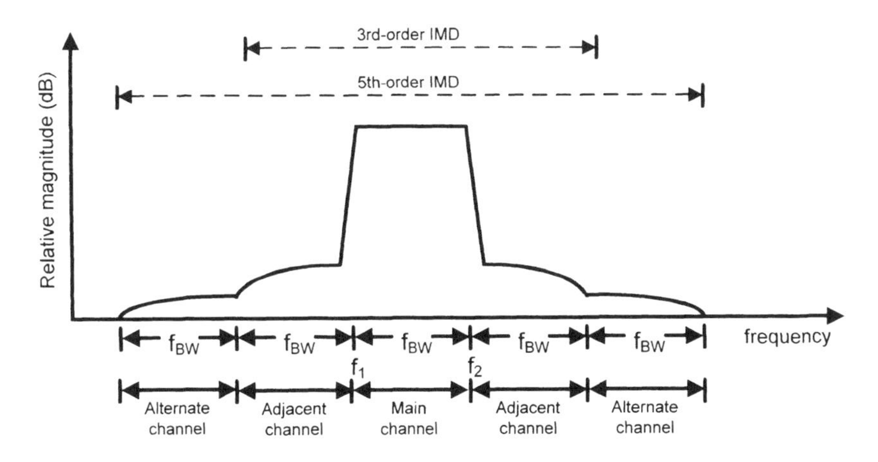

Figure 5.5. Typical spectrum of a digitally modulated RF signal showing spectral regrowth.

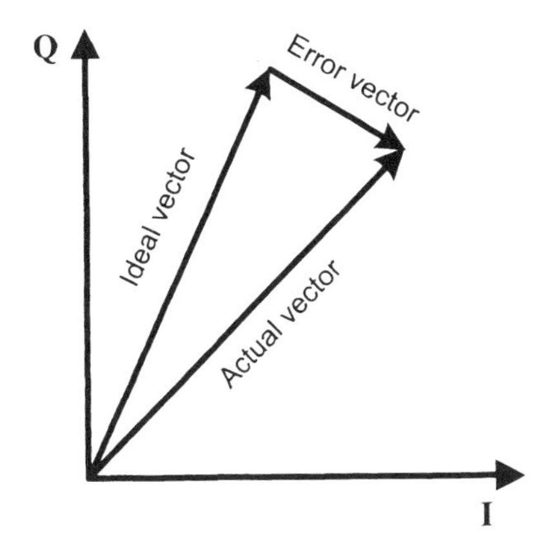

Figure 5.6. An illustration of the error vector for a digitally modulated signal.

Error vector magnitude (EVM) is a measure of the effect of nonlinearity on the accuracy of detection at the receiver. It quantifies the quality of the received digitally modulated signal. It is defined as the magnitude of the difference between the desired and the actual received signal vectors (see Figure 5.6), and is usually expressed as a fraction or percentage of the signal amplitude. Because it changes continuously, with every symbol transition, EVM is generally defined as the root-mean-square (RMS) value of the error vector over time. Sometimes the peak value is also used.

5.5 THERMAL INSTABILITY AND BALLASTING

A large-area power transistor used in a power amplifier may be regarded as a parallel combination of small-area transistors. In fact, in practice, power transistor cells are constructed in this manner. The distribution of current among the small transistors, or unit cells, that constitute the large power cell is an important consideration in RF power amplifier design, particularly in power amplifiers using bipolar transistors. To elaborate, consider a bipolar power transistor consisting of several small unit transistors, biased by the same V_{BE}. A unit transistor that has a temperature slightly higher than that of the others will carry more current than the others. If the temperature coefficient of the current is positive, that is, if the collector current increases with increasing junction temperature, this will increase its current further, resulting in a further increase in temperature, and so on. Thus, the positive thermal feedback mechanism will ultimately result in current crowding in a small area of the large power cell, irrespective of the magnitude of difference in initial temperature among the unit transistors. This instability of current distribution in the power transistor may initiate second breakdown. This is a phenomenon where, if a critical temperature is reached at a hotspot, an intrinsic zone is formed that short-circuits the space charge region of the P–N junction, resulting in a voltage reduction over the transistor. Current runaway due to thermal instability may even cause burnout of the transistor. One solution to overcome this is the use of the so-called ballast resistors. Resistances are placed at the emitters of the unit transistors, as shown in Figure 5.7a. These emitter ballast resistors, because of the voltage developed, would serve to oppose sizable buildup in current concentrations, and reduce the power cell's susceptibility to thermal runaway. Thus, ballast resistors help decrease the possibility of thermal runaway by reducing the nonuniformity of current distribution among the parallel unit cells in a power transistor. It is possible to analyze whether thermal runaway will occur, from the thermal dependence of current on junction temperature, and calculate the minimum value of ballast resistance required to prevent thermal runaway. This analysis and design of the ballast resistance is discussed in Chapter 6. It is important to note the mechanism of thermal failure differs for different types of devices. Silicon BJTs, for instance, have a positive feedback between the current and base–emitter junction temperature, that is, current gain (β) increases with temperature. Because of this positive thermal feedback mechanism, at high current densities, current crowding occurs, local hotspots are created, and this leads to thermal runaway and junction destruction [7]. In AlGaAs/GaAs HBTs, the current gain decreases with temperature. In this case, self-heating at high current densities, which results from the increase of the collector–emitter voltage (V_{CE}), leads to a thermal failure phenomenon called the *current gain collapse effect* [8]. Although the current gain of the SiGe HBT decreases with temperature, no current collapse occurs even at high collector–emitter bias voltages. Instead, at high current densities, the emitter current in a SiGe HBT may be constricted to localized hotspots, which can eventually lead to second breakdown, as in the case of Si BJTs [9].

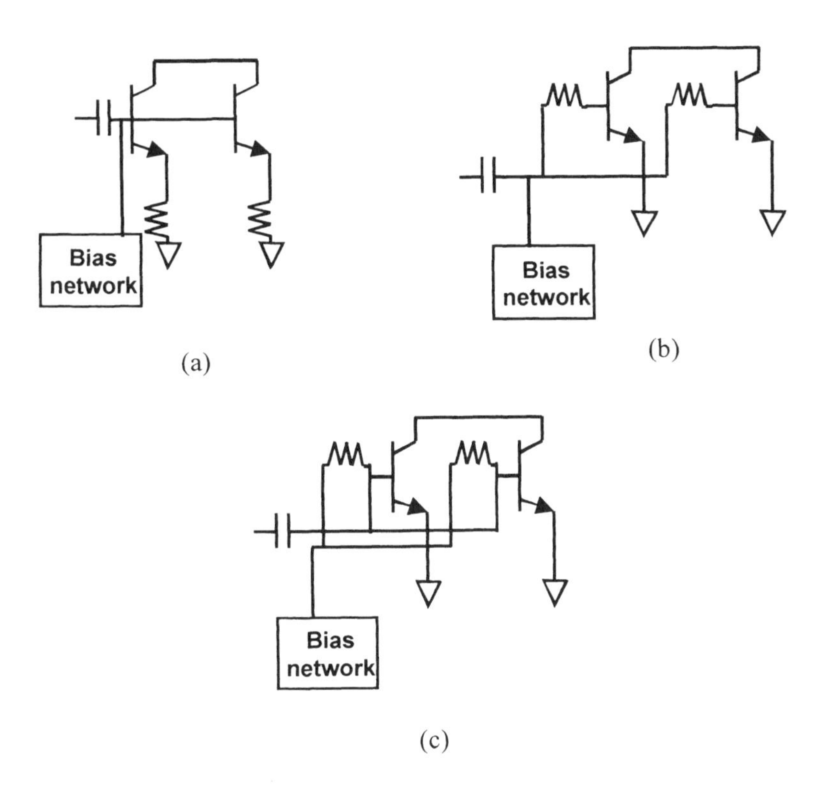

Figure 5.7. Ballasting schemes for bipolar power amplifier design: (a) emitter ballasting; (b) traditional base ballasting; and (c) "split" base ballasting.

Ballast resistors may be placed at the base terminal of the bipolar transistor, instead of the emitter. This is useful because emitter ballasting results in the loss of voltage headroom in low-voltage applications, apart from reducing gain. While the effective ballast resistance is small in a large power cell, and consequently this loss may not be significant, in smaller driver stages, it can result in a significant loss of gain and power. However, base ballasting cannot substitute emitter ballasting in all bipolar devices, owing to differences in the device failure mechanisms, as described above. Emitter ballasting resistors can be used in all types of bipolar transistors, but the use of base ballasting resistors requires a negative thermal feedback mechanism [10]. Therefore, base ballasting resistors are suitable for GaAs HBTs and SiGe HBTs, but not for Si BJTs. The value of the base ballast resistance to be used is β times the emitter ballast resistance value. There are different configurations possible, when using base ballasting. Figure 5.7b shows the traditional base ballasting scheme, where the base ballast resistor is placed in series with the RF input signal path. However, this scheme results in a rather large loss of RF power in the series resistance. To overcome this limitation, the ballast resistor can be removed from the RF path and placed only in the DC base bias path. Since the ballast resistor is used to provide thermal stability, and since the large constant current that is mainly responsible for causing instability is present in the DC path, this scheme effectively prevents thermal runaway without either consuming precious voltage swing or causing RF power loss. There are a couple of different possibilities in using this strategy. One involves placing a small capacitor in parallel with each base ballast resistor, to effectively short it out from the RF path. However, in some cases, the size of the unit transistor and its ballast resistor are such that adding a capacitance in parallel with each ballast resistor would result in a significant increase in the overall area occupied by the power transistor. The arrangement of Figure 5.6c can then be used, where the RF and DC base bias paths are "split." (Incidentally, the parallel capacitor scheme is also a "split" one, although the split is achieved in a slightly different way.) Since the ballast resistance is high in comparison with the device input impedance, the incoming RF signal is not lost into the bias network. This scheme enables the removal of the area-hungry RF-blocking inductor from the base bias path. Since the base ballast resistance is included only in the DC path, its value can be increased above the minimum estimated value, to incorporate a good margin to ensure thermal stability, without having to worry about RF power loss. Ballasting is not required in FET-based power amplifiers, because the thermal feedback mechanism that causes runaway in bipolar transistors is not present in FETs.

REFERENCES

1. H. L. Kraus, C. W. Bostian, and F. H. Raab, *Solid State Radio Engineering*, Wiley, New York; 1980.
2. F. H. Raab, Idealized operation of the class E tuned power amplifier, *IEEE Trans. Circuits Systems*, **CAS-24**(12): 725–735 (Dec. 1977).
3. N. O. Sokal and A. D. Sokal, Class E, A new class of high efficiency tuned single-ended power amplifiers, *IEEE J. Solid-State Circuits*, **SC-10**(3): 168–176 (June 1975).
4. F. H. Raab, Class-F power amplifiers with maximally flat waveforms, *IEEE Trans. Microwave Theory Tech.*, **45**(11): 2007–2011 (Nov. 1997).
5. N. Deshpande, S. Stanton, and M. Hurst, ACPR specs place demands on WCDMA base-station amplifiers, *Wireless Sys. Design*, pp. 15–21 (August 1999).
6. Agilent Technologies, Characterizing digitally modulated signals with CCDF curves, *application note – literature number 5968-6875E*, online, available at http://cp.literature.agilent.com/litweb/pdf/5968-6875E.pdf.
7. R. P. Arnold and D. S. Zoroclu, A quantitative study of emitter ballasting, *IEEE Trans. Electron Devices*, **21**(7): 385–391 (July 1974).
8. W. Liu, S. Nelson, D. Hill, and A. Khatibzadeh, Current gain collapse in microwave multifinger heterojunction bipolar transistors operated at very high power densities, *IEEE Trans. Electron Devices*, **40**(11): 1917–1927 (Nov. 1993).

9. J. Zhang, H. Jia, P. H. Tsien, and T. C. Lo, Emitter-ballasting-resistor-free SiGe microwave power heterojunction bipolar transistor, *IEEE Trans. Electron Devices*, **43**(2): 245–251 (Feb. 1996).
10. W. Liu, A. Khatibzadeh, J. Sweder, and H. F. Chau, The use of base ballasting to prevent the collapse of current gain in AlGaAs/GaAs heterojunction bipolar transistors, *IEEE Trans. Electron Devices*, **43**(2): 245–251 (Feb. 1996).

6

POWER AMPLIFIER DESIGN IN SILICON

6.1 INTRODUCTION

The growth of wireless communications, driven mostly by system-level considerations, has resulted in stringent demands being placed on power amplifier performance. The compelling need to develop cheaper and more efficient power amplifiers to satisfy these requirements has motivated more sophisticated power amplifier designs, where the power amplifier is accompanied by a host of ancillary circuits that perform support functions. Silicon CMOS is ideal for the implementation of such baseband control circuitry. Thus, silicon-based RF power amplifiers are attractive because they enable integration of the power amplifier and the control circuits on the same die. Further, they create the potential for integrating the power amplifier with other RF blocks on a single chip. A higher level of integration reduces overall cost, which makes a Silicon-based design preferable to a multichip III–V design. Silicon germanium (SiGe) BiCMOS technology, which combines Si MOSFETs with SiGe HBTs on the same silicon substrate, is an attractive choice for building a silicon-based power amplifier subsystem. The cutoff frequency (f_T), DC current gain (β), knee voltage, and other characteristics of the SiGe HBT offer considerable promise for RF power amplifier implementation [1]. In addition, the thermal conductivity of silicon (1.5 W/cm-°C) is 3 times higher than that of GaAs (0.49 W/cm-°C), resulting in better self-heating and current collapse characteristics. However, the low breakdown (maximum BV_{CEO} of 6 V) and Early voltages (~125 V) of SiGe HBTs compared to gallium arsenide (GaAs) HBTs, as well as the significantly higher parasitics, make power amplifier design with SiGe HBTs challenging.

SiGe HBTs offer the best characteristics for RF power amplifier design among silicon-based technologies. Design of power amplifiers in standard Si CMOS has also been the subject of active research [2]. In silicon-based power amplifier design, the focus is on improved and innovative design techniques, to overcome the inherent limitations of the device technology. This chapter will serve to illustrate the challenges in designing power amplifiers in silicon, and will discuss design techniques to overcome these challenges, using practical examples of SiGe HBT and CMOS power amplifier design. First, the design of a 2.4-GHz SiGe HBT-based power amplifier module is described. The procedure for designing the power amplifier, and the various factors to be taken into account during the design process, are discussed in detail. The design techniques outlined here are applicable to any power amplifier design in general. Additionally, this design illustrates the integration of semiconductor technology with high-performance passives to achieve high power-added efficiency. Further, the use of a novel technique to adjust the active device area is illustrated as a means to achieve both higher efficiency, and improvement in linearity, in silicon-based power amplifiers. The use of this technique for both SiGe HBT and CMOS power amplifiers is described. This will also serve to introduce efficiency enhancement techniques in linear power amplifiers, which is the subject of the next chapter.

6.2 A 2.4-GHz High-Efficiency SiGe HBT Power Amplifier

The 2.4-GHz ISM band is very important in digital wireless communications, with several applications such as Bluetooth, Home RF, DECT, wireless local loop, and IEEE 802.11b/g wireless LAN systems. SiGe HBT power amplifiers have been reported for many applications, such as GSM, DECT, AMPS, and CDMA [3–6]. In this section, a 2.4-GHz SiGe HBT-based power amplifier module that delivers up to 27.5 dBm output power, with a maximum power-added efficiency of nearly 47% [7], is described. Excellent harmonic suppression is achieved by a harmonic suppression filter at the output, which, along with the output match network, is implemented in a multilayer low-temperature co-fired ceramic (LTCC) substrate, on which the SiGe HBT power amplifier is mounted. Integration of the silicon power amplifier IC with high-performance passives on LTCC enables the module to achieve GaAs-like power and efficiency performance.

The use of a filter to suppress harmonics is quite common in power amplifier design. Most high-efficiency power amplifiers achieve high efficiency by operating under several decibels of compression, and therefore have a very high harmonic content in their output power spectrum, making it necessary to use a harmonic suppression mechanism. Harmonic suppression filters are widely used for this purpose. Various types of lowpass filters suitable for use in harmonic suppression, such as, open-stub and stepped-impedance filters [8,9], have been reported. However, all these have limitations that make it difficult to meet the requirements of modern communication systems. In the case of the open-stub lowpass filter, for instance, its distributed nature results in a large size and a narrow stopband. For sharp cutoff, the order of the stepped-impedance lowpass filter must be very high. This results in a large circuit and large insertion loss. The snake-type lowpass filter [10] can have attenuation poles at finite frequencies. However, these poles cannot be located near the passband, since the side-coupled capacitor in the shunt capacitor–inductor pair is too small. Therefore, the capacity to reduce the filter order is limited. In the example illustrated here, a compact filter with excellent harmonic suppression performance is implemented completely in an LTCC substrate. No external lumped components are used, and the output match network is also implemented in the LTCC substrate.

6.2.1 Circuit Design Considerations

The power amplifier discussed here is designed using SiGe HBT unit cells with an emitter area of 20 μm^2 (0.5 × 20 μm × two fingers). The devices used are the high breakdown devices (BV_{CEO} = 5.5 V) available in the process. The schematic of the designed power amplifier is shown in Figure 6.1. The portion of the schematic enclosed in the box represents the part of the power amplifier on chip. The rest is implemented in the LTCC substrate. It consists of three stages (to obtain the high amount of gain, 30 dB): two driver stages and a power stage; interstage, input and output match networks; and bias circuits for the three stages.

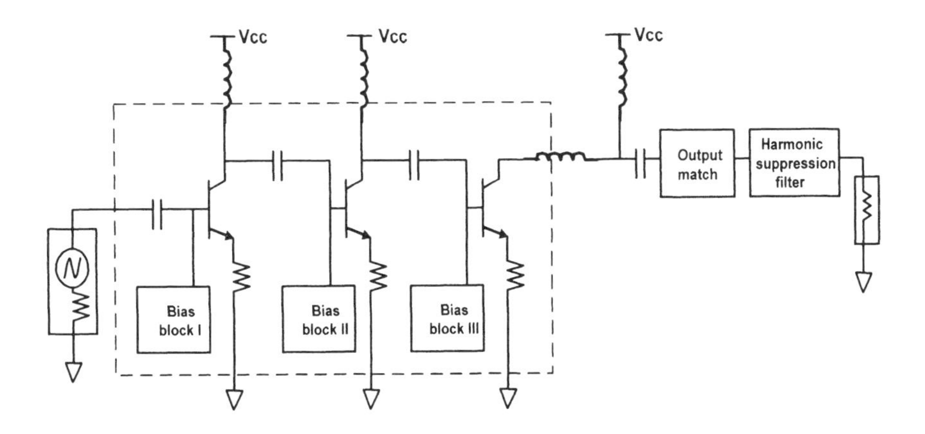

Figure 6.1. Simplified schematic of the SiGe HBT power amplifier.

The first step in designing the power amplifier circuit is the sizing of the total transistor periphery to handle the required power. Assuming a target power P of 750 mW (including a substantial margin over the minimum target of 500 mW), a knee voltage V_K of 0.5 V, and a nominal supply voltage of 3.3 V, the required collector bias current I_C can be calculated from load-line principles using

$$I_C = \frac{2P}{V_{CC} - V_K} \tag{6.1}$$

which results in an I_C of about 536 mA. Since the peak f_T of the device is at a current level of 0.35 mA/μm^2 (from the process data), beyond which the RF performance degrades because of the Kirk effect, this results in an emitter size of about 1530 μm^2. This translates to 76 unit transistor cells. In this design, a power transistor with 72 unit cells (emitter area = 1440 μm^2) was utilized, to enable easy arrangement of the transistor cells in the power stage.

The optimum load-line output impedance R_L to achieve the required output power is given by

$$R_L = \frac{V_{CC} - V_K}{I_C} \tag{6.2}$$

which results in a value of 5 Ω for R_L. The output match network in LTCC is designed to match this impedance to 50 Ω. To account for the device and parasitic capacitances that must be canceled out, a provision was made to mount a few surface-mount lumped components on the board. These could be subsumed in the embedded LTCC matching network in a later design iteration, if desired.

The procedure for sizing the driver stages is similar to that followed for the output power stage. The output power level of a driver stage is lower than that of the following stage by the estimated gain of that stage. The gain of the stages was estimated from simulations, with the addition of margins to account for parasitic losses (in addition to the estimated parasitics that were added in the simulation platform). The first and second driver stages were accordingly sized to be 120 μm^2 and 240 μm^2, respectively. The driver stages are small compared to the output power stage, since the output power demanded of these stages is relatively low. Further, most of the burden of the total gain is on these stages, and a smaller size has lesser parasitic losses, which helps maximize the gain of the amplifier stage.

Interstage matching is compact and is achieved using only on-chip capacitors, and bond wire and package lead inductances that also serve as RF chokes for the first two stages. Since bond wire inductances play a crucial role in matching, it is essential to analyze their properties and their effect on the circuit design. Bond wires are preferred over on-chip spiral inductors in many RF ICs because of their higher Q values. These higher Q values are essentially because bond wires have much more surface area per unit length compared to planar spirals, and hence less resistive loss. Also, they may be placed well above any conductive planes to reduce parasitic capacitances (thereby increasing the self-resonant frequency) and decrease the loss due to induced currents. The DC self-inductance, L, of a bond wire is given by [11]

$$L \cong \frac{\mu_0 l}{2\pi}\left(\ln\frac{2l}{r} - 0.75\right) \tag{6.3}$$

where l is the length of the bond wire and r is its radius. This results in a thumbrule of approximately 1 nH/mm length for a standard bond wire of 1 mil diameter with length in the 1–2 mm range. The resistance of the standard bond wire is estimated to be about 125 mΩ/mm at 1 GHz [12]. Bond wires were used in this design extensively. Since a silicon-based process does not usually have through via holes, the emitters of the transistors have to be grounded externally via bond wires. These bond wires were "downbonded" to the metal paddle on which the IC is mounted (in the package) to reduce their length and hence minimize loss of gain and parasitic effects. A large number of closely spaced bond wires are used in parallel for downbonding and for the various pins, for instance, the RF output and the collector node pins, to reduce the effective inductance. Thus, it becomes essential to examine the mutual inductance coupling between the bond wires. For two bond wires of equal length l and radius r separated by a distance D, the mutual inductance M is given by

$$M \cong \frac{\mu_0 l}{2\pi}\left(\ln\frac{2l}{D} - 1 + \frac{D}{l}\right) \tag{6.4}$$

For a 1 mm length and a spacing of 0.2 mm (typical in an IC), the mutual inductance works out to be 0.3 nH, which represents a coupling coefficient of 30%. Further, the logarithmic dependence of M on spacing means that the coupling decreases rather slowly with distance. Thus, the mutual inductance between bond wires is significant and has to be accounted for while designing the power amplifier. Finally, the bond wire inductances and mutual inductances are not well controlled, because of possible variations in length and spacing in actual implementation, and therefore, the sensitivity of the design to variations in these values must be examined. In this design, the bond wires and package leads were modeled by lumped inductances, with the mutual inductance effect incorporated as an increase in the effective inductance, for simplicity. The effect of variation in the inductances was also examined by extensive simulation to ensure that sufficient margin was included in the design to offset these variations.

The power amplifier is packaged in an 8-pin small-outline IC (SOIC) package. Package parasitics were estimated and included in the simulation test bench during the design of the amplifier. A bipolar current source based circuit was used to bias the transistor stages. The bias circuit was designed to minimize the variation in bias current to less 5% over a temperature range of 25–125°C. The output match network and harmonic filter are off-chip and implemented in LTCC. Figure 6.2 is a photograph of the designed power amplifier IC. It is about 1.2×0.9 mm in size.

Figure 6.2. Microphotograph of the SiGe HBT power amplifier IC.

6.2.2 Analysis of Ballasting for SiGe HBT Power Amplifiers

One important consideration in bipolar power amplifier design is ballasting, as discussed in Chapter 5. Ballasting is essential because any small mismatch in temperature between devices or emitter fingers in parallel, would trigger thermal runaway due to the strong positive feedback between collector current and temperature, resulting in catastrophic destruction of the device [13,14]. Ballasting ameliorates this thermal instability by making the current and temperature distribution more uniform. To begin with, given the good thermal conductivity of silicon compared to GaAs, it needs to be ascertained whether ballasting is required for SiGe HBT power amplifiers. This can be analyzed starting with the following dependence of collector current on junction temperature [15]

$$I_c = I_{so}.\exp\left(\frac{qV_{be} - qI_c\left[\frac{r_e + R_{eb}}{\alpha} + \frac{r_b}{\beta}\right] - E_g}{kT}\right) \tag{6.5}$$

where R_{eb} is the emitter ballast resistance per unit transistor cell, and temperature T is given by

$$T = T_A + R_{th}\, I_c\, V_{cc} \tag{6.6}$$

where R_{th}, the thermal resistance of the device is given by [16]

$$R_{th} = \frac{\ln(4L/W)}{\pi k L} \tag{6.7}$$

where L and W are the length and width of the emitter of the device, respectively, and k, the thermal conductivity of silicon, is ~0.1 mW/μm-°C at 350°C. From this, R_{th} is computed to be ~0.81°C/mW. The highest operating ambient temperature considered in this design is 85°C, and taking into account the effect of the thermal resistance of the package and the die, the highest value of T_A is found to be 125°C. Assuming a low Germanium content in the base, $E_g(T)$ can be approximated by the temperature dependence of E_g of Si [17]

$$E_g(T) = E_{g0} - 3.6 \times 10^{-4}\, T \tag{6.8}$$

where E_{g0} is 1.21 eV, and the temperature dependence of β can be represented by

$$\beta(T) = \beta(T_A)\left(\frac{T}{T_A}\right)^{XTB} \tag{6.9}$$

where T_A is 298K and $XTB = -0.7$ for the technology under consideration. The threshold for onset of thermal runaway is given by

$$\left(\frac{\partial J_c}{\partial V_{be}}\right)^{-1} = 0 \tag{6.10}$$

Figure 6.3 shows J_c as a function of V_{be} for different values of R_{eb}. It is clearly seen that if no ballasting is present ($R_{eb} = 0$), thermal runaway will occur. Keeping in mind the approximations involved in the above analysis, being conservative, a value of R_{eb} of 3 Ω (per unit transistor cell) is used in this design.

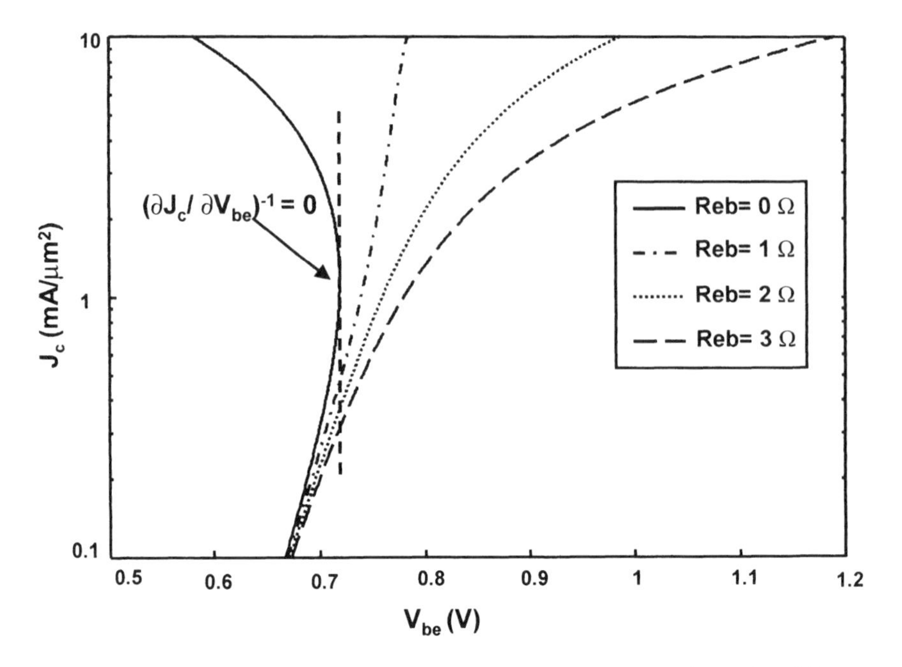

Figure 6.3. Collector current density versus base–emitter bias voltage for different emitter ballast resistance values.

6.2.3 Harmonic Suppression Filter and Output Match Network

The harmonic suppression filter and output match network are designed in a 10-metal-layer LTCC process. Figure 6.4 shows the designed filter and output match network. The filter consists of a transmission line in parallel with a capacitor C. The attenuation poles of this filter are determined by the zeros of Equation 6.11 [18]

$$S_{21}, S_{12} = \frac{2jY_0[\omega C - Y_c \csc(\beta l)]}{\Delta Y} \tag{6.11}$$

$$\Delta Y = Y_0^2 + Y_c^2 + 2jY_0[\omega C - Y_c \cot(\beta l)] + \omega C Y_c[\cot(\beta l) - \csc(\beta l)]$$

where Y_0 is the characteristic admittance of the input and output port, Y_c is that of the transmission line, l is its length, and β is its propagation constant. The filter is designed to give deep suppression of the second and third harmonics. A $\lambda/4$ length RF-shorted stub is used to achieve a short at the second harmonic. The output match network, which is also implemented in LTCC using transmission lines, tuned in conjunction with the harmonic suppression filter, presents nearly a short at the second harmonic and nearly an open at the third harmonic. The harmonic suppression performance of the filter is shown in Figure 6.5. As can be seen, the suppression of the second and third harmonics is better than 45 dB. The filter exhibits a loss of about 1 dB at the fundamental frequency (2.4 GHz).

A few comments on evaluating the performance of the LTCC filter–output match network are appropriate here. The network presents the optimum impedance desired by the power amplifier at its input and transforms this impedance to 50 Ω at its output. However, the network analyzer, which is used to measure the *S*-parameters of the network, is a 50-Ω system. Therefore, the network analyzer-measured *S*-parameters were transformed into those seen by a 5-Ω (the designed input match) port on the input side and a 50-Ω one on the output side, as illustrated in Figure 6.6. The resulting S_{11} parameter is plotted on a Smith chart referenced to 5 Ω (i.e., the center of the Smith chart corresponds to 5 Ω) in Figure 6.7. This plot clearly illustrates the multiple functions performed by the designed network. At the fundamental frequency (f_0 = 2.4 GHz), the network presents the desired optimum impedance to the power amplifier; it is nearly a short at the second-harmonic frequency and nearly an open at the third-harmonic frequency. This harmonic control behavior results in waveform shaping similar to that in Class F (although the power amplifier is not biased at cutoff as in a traditional Class F amplifier), and contributes to improving the output power and efficiency of the amplifier, enabling the module to achieve GaAs-like performance. (See Figure 6.8 for simulated current and voltage waveforms at the output of the third stage.)

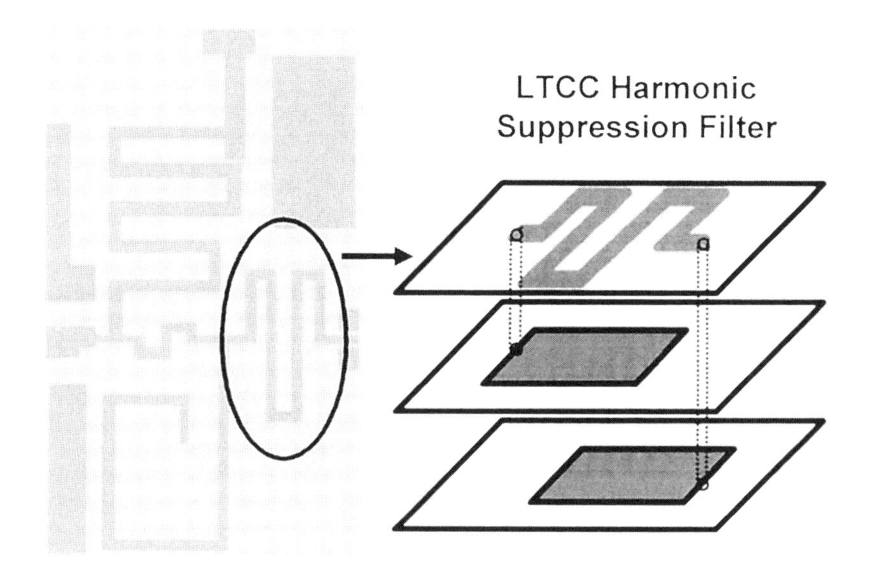

Figure 6.4. LTCC harmonic suppression filter and output match network.

6.2.4 Performance of the Power Amplifier Module

Figure 6.9 is a photograph of the designed power amplifier module consisting of the SiGe HBT power amplifier IC mounted on an LTCC substrate containing the output network. The module is very compact, and is significantly less than 0.5 in^2 in area. The output power and efficiency of the power amplifier with the LTCC harmonic suppression filter is shown in Figure 6.10. The power amplifier achieves an output power of 27 dBm at 0 dBm input with a PAE of 45% at 2.4 GHz and supply voltage V_{CC} = 3.3 V. At 5 dBm input, the output power is 27.5 dBm with a PAE of 47%. The linear gain of the power amplifier is 35 dB. The second and third harmonics are –44 dBc and –49 dBc, respectively, at 0 dBm input. The sharp rolloff of the output filter characteristic also results in the suppression of other out-of-band spurious signals.

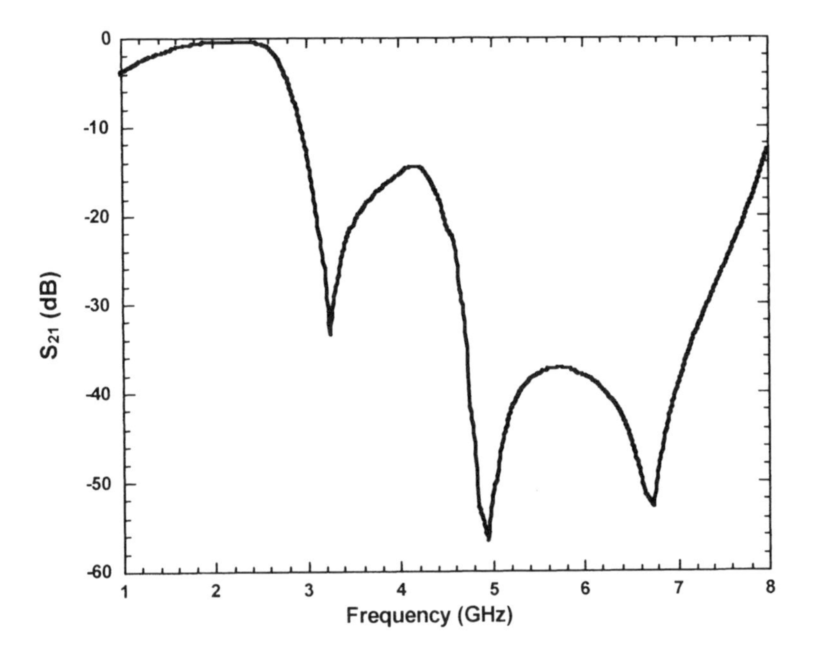

Figure 6.5. Harmonic suppression performance of the filter.

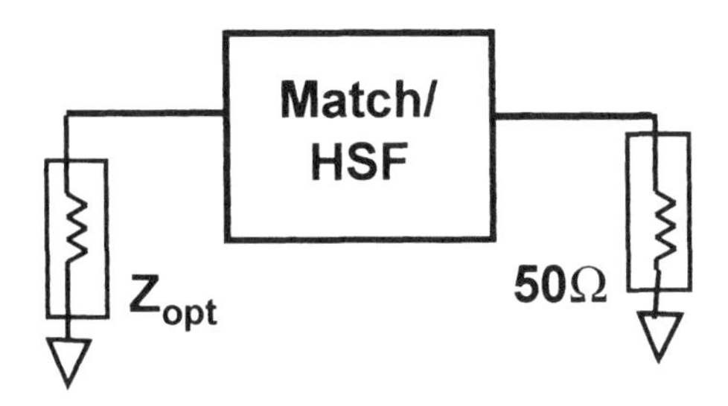

Figure 6.6. Testbench for proper evaluation of the LTCC output network.

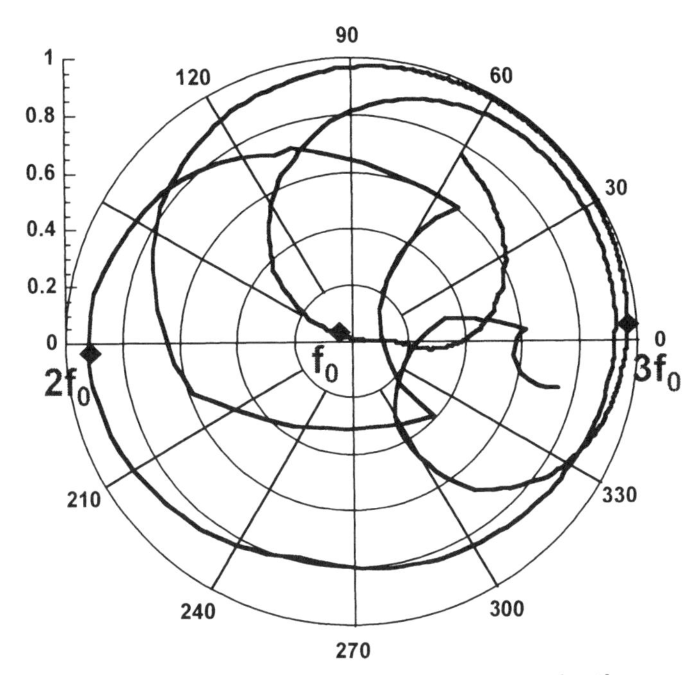

Figure 6.7. S_{11} of the output network with the configuration of Figure 6.6.

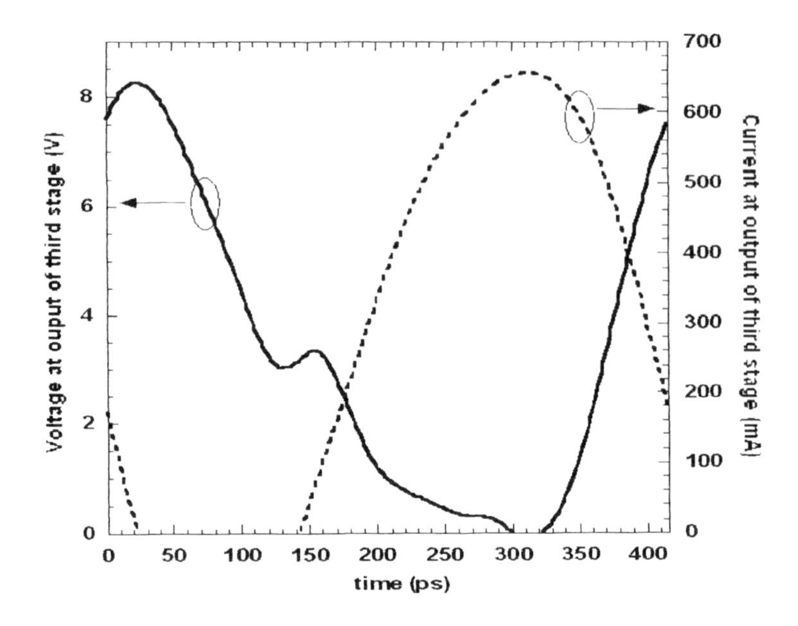

Figure 6.8. Voltage and current waveforms at the output of the third stage.

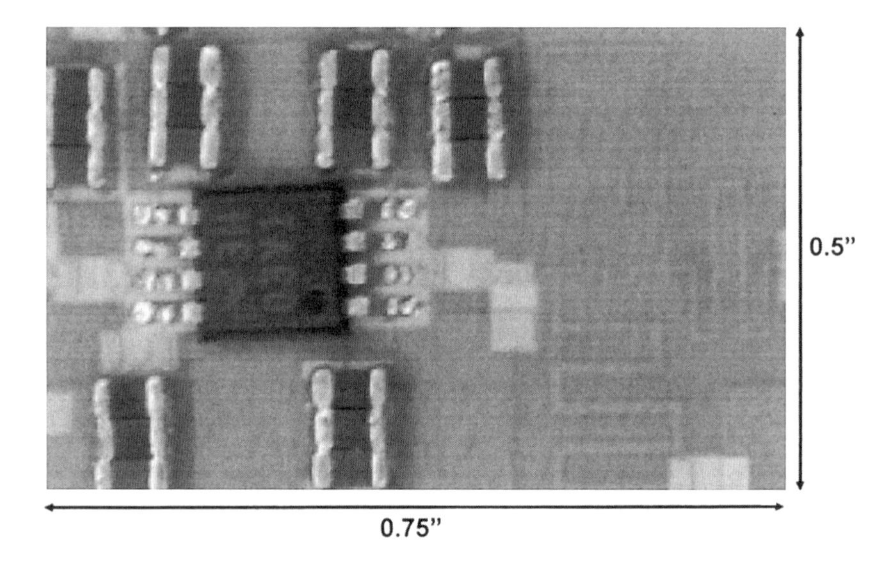

Figure 6.9. The 2.4-GHz SiGe HBT power amplifier module.

6.3 RF POWER AMPLIFIER DESIGN USING DEVICE PERIPHERY ADJUSTMENT

In current mobile communication standards, such as, CDMA and WCDMA, the output power of a handset must be reduced when the interference level to nearby channels is high. This is done by reducing the amplitude of the signal generated by the variable-gain amplifier (VGA) prior to power amplification. The efficiency of a power amplifier at this low input power is much less than that at high power level. Therefore, the power amplifier must be designed to be able to control DC consumption, to improve the conversion efficiency.

A few techniques have been developed for improving low-power efficiency in single-chip power amplifier ICs. The switching of collector or drain quiescent current level is widely used in commercial power amplifier monolithic microwave integrated circuits (MMICs) to reduce DC power consumption, while keeping the linearity level within the system requirements. This can increase the average power-added efficiency (PAE) of the power amplifier, and prolong the battery life of a mobile handset, with the tradeoff of degradation in other aspects of performance, such as gain and linearity. Using this technique, limited PAE improvement has been observed in Class AB amplifiers where the quiescent current in the low-power mode is small. This technique will be more effective in Class A amplifiers where the quiescent current is usually high (see Figure 6.10).

Alternatively, a technique described as "stage-bypassing" has been proposed to improve efficiency when operating at low power [19,20]. This technique uses a switch to bypass the output stage in low-power mode, as shown in Figure 6.11a, and shuts down the output transistor, which decreases the overall DC power consumption and increases the PAE. Limitations of this approach are the use of a switch, which introduces loss, and the gain difference between the modes of operation. An external low-loss switch can be used; however, the extra component adds cost and increases area.

Figure 6.11b illustrates an approach that improves efficiency by adjusting the active device area [21–23]. Improvement in PAE can be achieved at the cost of increased complexity in matching network design. This technique has provided the best improvement in PAE while meeting the specifications of current wireless communication standards, such as IS-95 and WCDMA.

In this section, the active device area adjustment technique is illustrated for silicon-based designs, in both standard digital CMOS, and SiGe HBT. Further, a linearization method using the cancellation of third-order intermodulation terms of parallel FETs, known as *derivative superposition* [24], is implemented to enhance the linearity of the CMOS power amplifier. These examples illustrate the use of advanced design techniques to overcome silicon device technology limitations to achieve linearity and efficiency performance suitable for practical wireless communication systems.

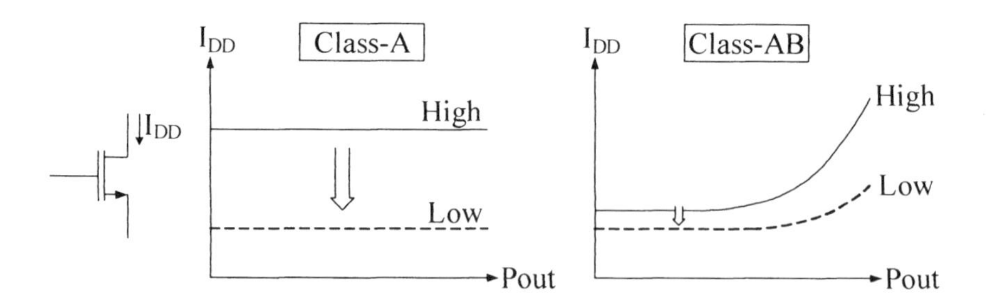

Figure 6.10. Reducing the bias current to improve power amplifier efficiency at low power.

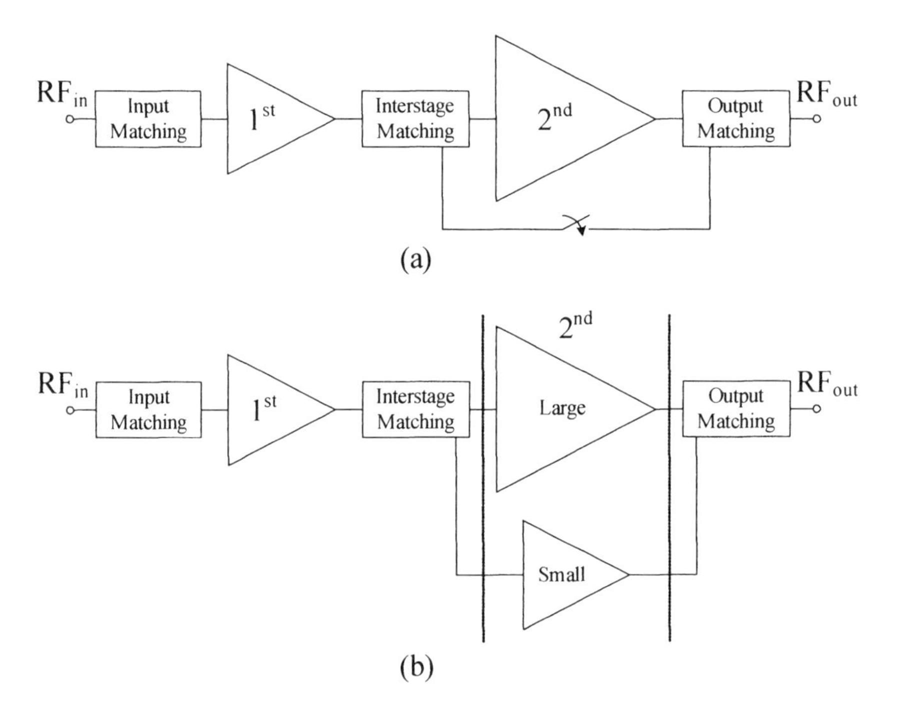

Figure 6.11. Topologies for efficiency enhancement of single-chip linear power amplifiers: (a) stage bypassing; and (b) active area adjustment.

6.3.1 Analysis of the Device Periphery Adjustment Technique

The device periphery adjustment technique aims to adjust device periphery to reduce DC consumption, without sacrificing device performance. In small-signal circuits, figures of merit such as cutoff frequency f_T, the frequency at which the short-circuit current gain equals one, or maximum oscillation frequency f_{max}, the frequency at which the conjugate matched power gain (maximum available power gain) equals one, are used as figures of merit for the performance of a transistor. However, it is difficult to find a similar figure of merit for power amplifier design, since the power amplifier is matched for maximum output power. Roughly, f_{max} may be used in the driver stage, as the signal level is not too high and the matching design usually assumes a conjugate-matched condition. In the output stage, load-pull measurements or simulations are used in practice, to find a suitable device for a particular output power and bias level. For this design, f_{max} is used initially as a figure of merit and load-pull measurements are then used to determine the device size for the desired output power and efficiency.

f_{max} of a MOSFET is given by

$$f_{max} = \sqrt{\frac{f_T}{2\pi R_g C_{gd}}} \tag{6.12}$$

where R_g is the gate resistance and C_{gd} is the gate–drain capacitance. Usually f_{max} is larger than f_T since R_g can be reduced greatly by careful layout techniques and silicided gate. f_{max} profiles of MOSFET and SiGe HBT devices used in this work are shown in Figures 6.12 and 6.13, respectively. For bipolar transistors, base resistance R_b and base–collector capacitance C_{bc} are used instead of R_g and C_{gd}. A transistor exhibits a peak in f_{max} only over a small range of bias current. It is advisable to maintain the device current density at peak f_{max} to maximize device performance. Therefore, to reduce the DC current for power amplifiers operating in low-power mode, lowering the device periphery is a better option than adjusting the DC current of the same device.

The next step is to realize this concept in MOSFET and SiGe HBT power amplifier designs. In the case of the MOSFET, the design has two separate FET devices with different W/L ratios, connected in parallel at the output stage, as shown in Figure 6.14. The device with the smaller W/L ratio is intended for operation in the low-power mode, whereas the larger device is operated along with the smaller one in the high-power mode. This reduces DC power consumption in the low-power mode, without reducing device performance metrics, such as g_m and f_{max}, since the device current densities are unaltered. Each transistor is terminated with an optimum output impedance termination for maximizing efficiency at the desired output power. To account for the effects of pad parasitics and bond wires, load-pull measurements are performed to obtain realistic matching conditions. The same design method is used in the case of the SiGe HBT power amplifier.

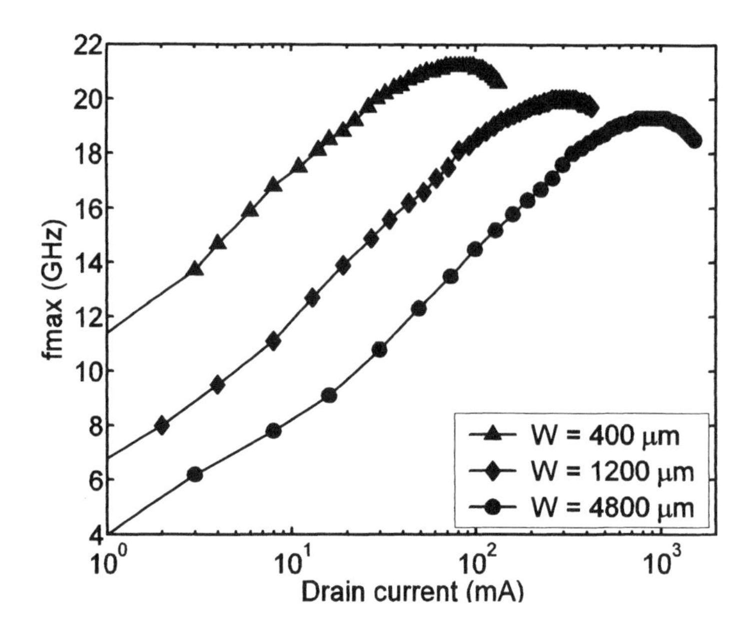

Figure 6.12. f_{max} of MOSFET devices with different total width. These results include the effect of pad and interconnect parasitics (device length $L = 0.4$ μm).

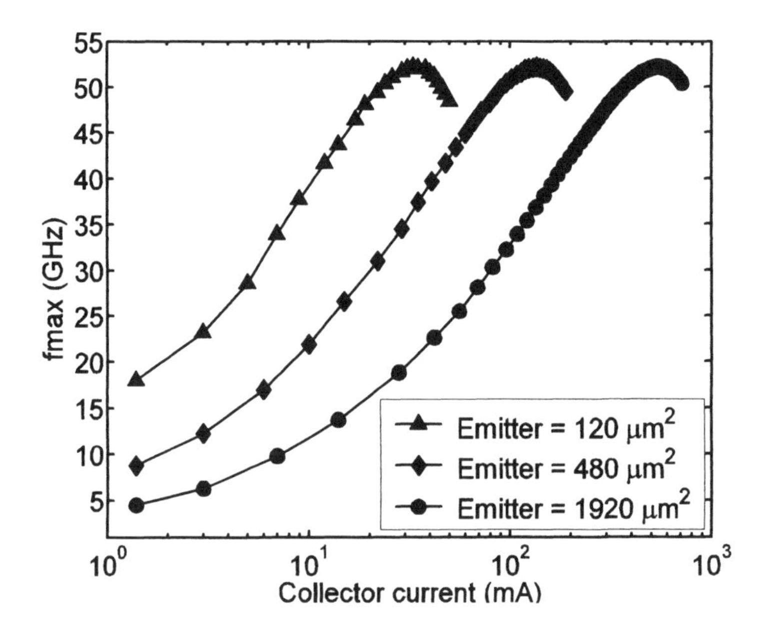

Figure 6.13. f_{max} of SiGe HBT devices with different total emitter area (pad and interconnect parasitics are not included).

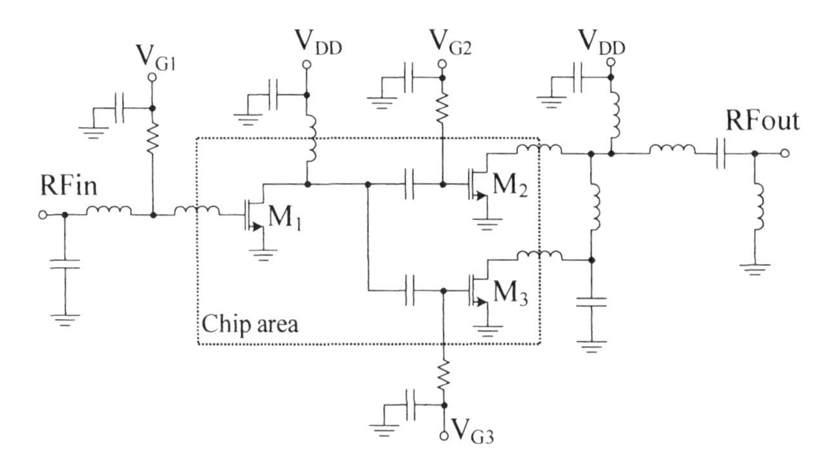

Figure 6.14. Simplified schematic of the CMOS power amplifier with two-step power control (*W*/*L*: M_1 = 1200 μm/0.4 μm; M_2 = 4800 μm/0.4 μm; M_3 = 1200 μm/0.4 μm).

A few practical concerns are the phase difference between two signal paths (one path to the output is through the large transistor, and the other is through the smaller transistor connected in parallel), signal loss due to the loading of devices in parallel, and stability issues. The interstage and output match networks need to transform optimum impedances to each transistor as well as minimize the phase difference, which causes distortion or signal cancellation. The output impedance of the output-stage devices must be taken into account when designing the output match network, as it becomes a load to the other transistor. Also, the circuit must be properly grounded and bypassed since it is prone to oscillation due to the parasitics from additional parallel devices.

6.3.2 1.9-GHz CMOS Power Amplifier

The power amplifier (schematic shown in Figure 6.14) is designed for operation at 1.9 GHz, using a standard 0.18-μm CMOS process. Thick-oxide devices (L = 0.4 μm) are used because of their higher drain–source breakdown voltage. A drain bias voltage (V_{DD}) of 2.4 V is used for both stages. The driver-stage transistor (M_1) is designed to have a total gate width of 1200 μm, while the output-stage transistors are sized at 4800 μm (M_2) and 1200 μm (M_3), respectively. The total die area is approximately 0.55 × 0.7 mm^2, including pads. A microphotograph of the IC is shown in Figure 6.15, and the PCB board with the IC mounted on it, is shown in Figure 6.16. The matching networks are off-chip, except for the interstage capacitors which are on-chip poly–poly capacitors. Load-pull measurements were used to find optimum matching impedances for each transistor. All inductors are realized using bond wires, trace lines, and off-chip SMT inductors, to minimize loss. A detailed schematic, including the bias networks and component values, is shown in Figure 6.17.

The bias network must be designed carefully to avoid oscillations. Silicon-based RF power amplifiers are more susceptible to instability and oscillation, because of the lower substrate resistivity and higher parasitic capacitance, compared to other technologies such as GaAs HBTs and MESFETs. Parasitic oscillation normally occurs in the low-frequency region (kHz to a few hundred MHz range) where the gain of the circuit is high. Therefore, it is important to avoid the forming of a low-frequency tank, which normally occurs around the RF chokes. In this design, low-value RF inductors (realized by transmission lines TL_2 and TL_3) are connected to the drain of the transistors and then bypassed by small capacitors (22 pF). These inductors present an impedance of around 5–8 times higher than the impedance transferred from the matching networks, which is enough to block the RF signal. This inductance is low enough to prevent resonance with the parasitic capacitance at the drain node in the low-frequency range. Then, higher value RF chokes can be connected to provide more isolation between RF and DC. Large capacitors (1 nF and 10 μF) are used for bypassing low frequency signals from external circuits and creating low-baseband-impedance paths looking back from the power amplifier circuit to the biasing networks, which help minimize intermodulation (IMD) asymmetry [25,26].

Figure 6.15. Chip photograph of the 1.9-GHz CMOS power amplifier.

The output match network transforms the 50 Ω load down to (5 – j5) Ω, which is the optimum load for the 4800-μm device (M_2) to achieve 23 dBm output power at 1-dB compression (P_{1dB}). An additional L-network is connected from the main output match network to create an optimum load of (8 – j10) Ω for M_3, which is designed to deliver 17 dBm of output power at P_{1dB}. The output impedance of each transistor in the output stage becomes an undesirable load to the other transistor, and reduces the output power to the 50-Ω load. The output impedance of M_2 is found to be (7 – j8) Ω from the device model and greatly affects the matching network design for M_3. The output impedance of M_3 has a much higher value than the optimum load for M_2, thereby resulting in much less loading effect.

The CMOS power amplifier is designed to have a P_{1dB} of 23 dBm at 1.9 GHz. The gain of the amplifier in low-power and high-power modes are approximately 16 and 21 dB, respectively, as shown in Figure 6.18. PAE in the low-power region, shown in Figure 6.19, is substantially improved as a result of reducing the DC power consumption while providing good match at low power level. The PAE in the low-power mode at 6 dB backoff is about 15%, which is an improvement by a factor of 2, compared to a normal Class AB amplifier (with all transistors in the design operating, as in the high-power mode), while the PAE improvement is a factor of > 2.5 at lower power levels. The IM3 of the power amplifier at low and high power levels is shown in Figure 6.20.

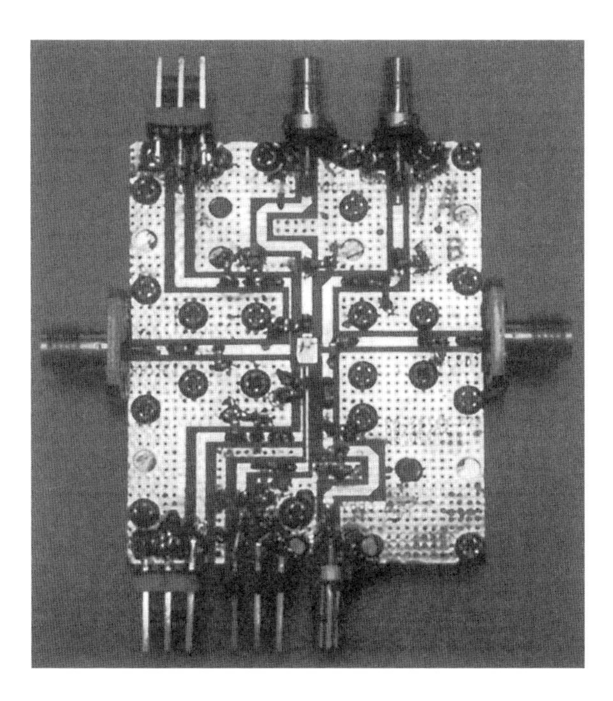

Figure 6.16. PCB board for the 1.9-GHz CMOS power amplifier.

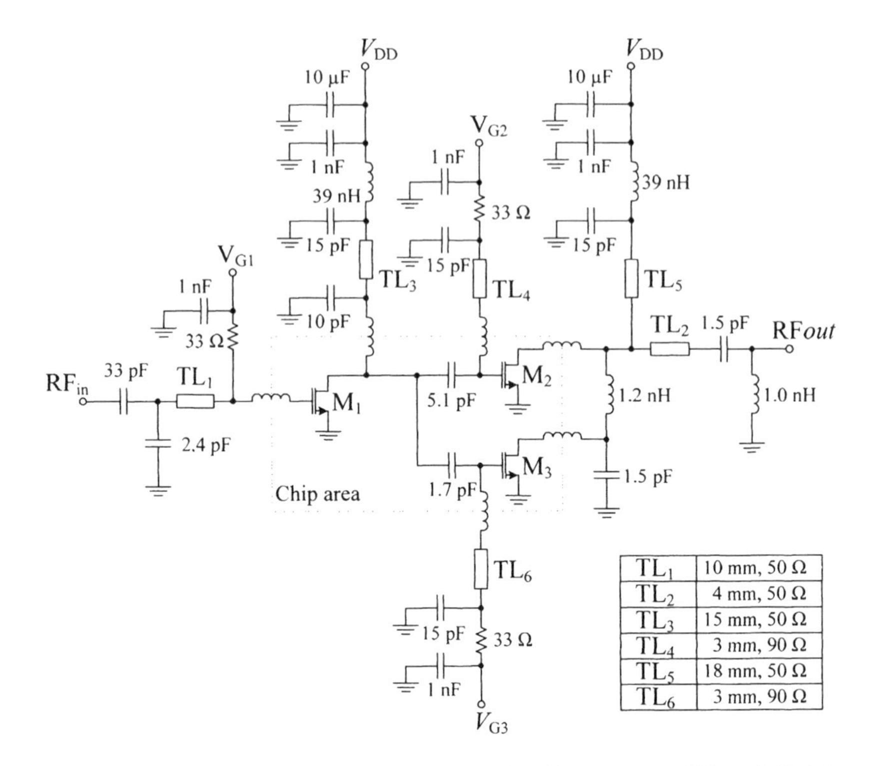

TL_1	10 mm, 50 Ω
TL_2	4 mm, 50 Ω
TL_3	15 mm, 50 Ω
TL_4	3 mm, 90 Ω
TL_5	18 mm, 50 Ω
TL_6	3 mm, 90 Ω

Figure 6.17. Detailed schematic of the 1.9-GHz CMOS power amplifier. (W/L: M_1 = 1200 μm/0.4 μm; M_2 = 4800 μm/0.4 μm; M_3 = 1200 μm/0.4 μm).

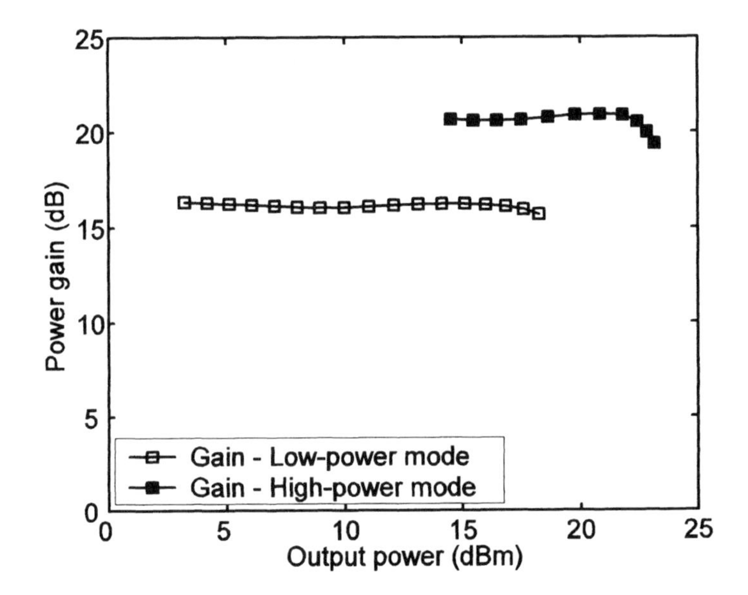

Figure 6.18. Power gain of the CMOS power amplifier at 1.9 GHz.

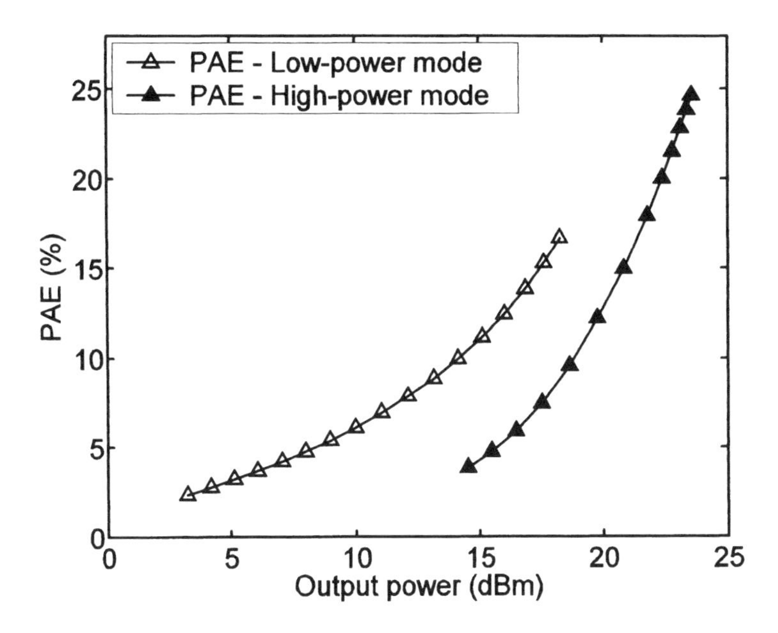

Figure 6.19. Power-added efficiency of the CMOS power amplifier at 1.9 GHz.

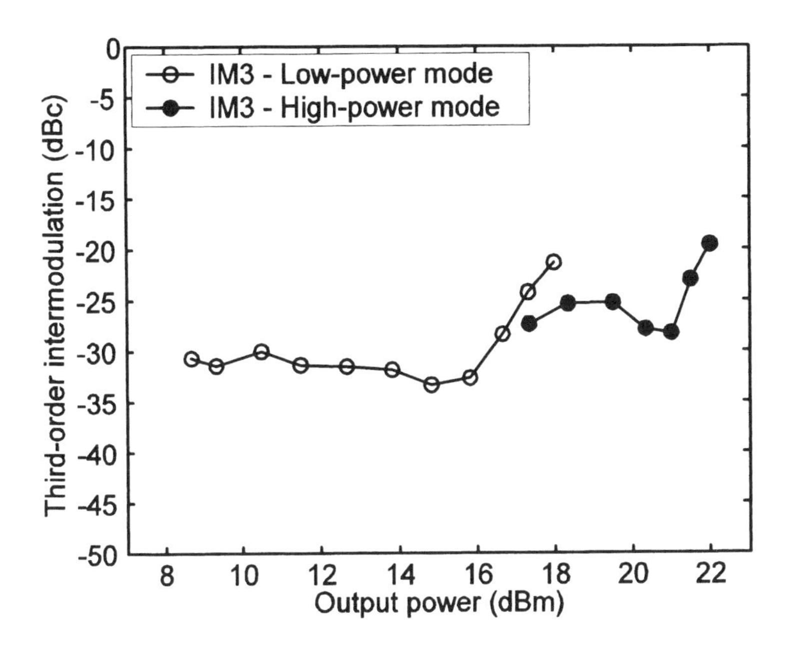

Figure 6.20. Third-order intermodulation distortion of the CMOS power amplifier with two-tone input signals at 1.88 and 1.8801 GHz (100 kHz signal separation).

6.3.3 1.9-GHz CDMA/PCS SiGe HBT Power Amplifier

In this section, the design of a 1.9-GHz SiGe HBT power amplifier for CDMA/PCS, based on the device periphery adjustment concept, is described. The design consists of two stages, with on-chip bias networks. The emitter area of the first stage is 480 μm^2. The second stage has two transistors with emitter areas of 480 μm^2 and 1920 μm^2, respectively. The input and output match networks are mostly off-chip, and use lumped components and transmission lines. Interstage matching capacitors and output-match capacitors of the smaller device in the second stage are realized on-chip using metal–insulator–metal (MIM) capacitors. A detailed schematic is shown in Figure 6.21. The total die size is about 1 mm^2 and the IC is packaged in a 4 × 4 mm 16-pin leadless package. The die photograph and prototype board layout are shown in Figures 6.22 and 6.23, respectively.

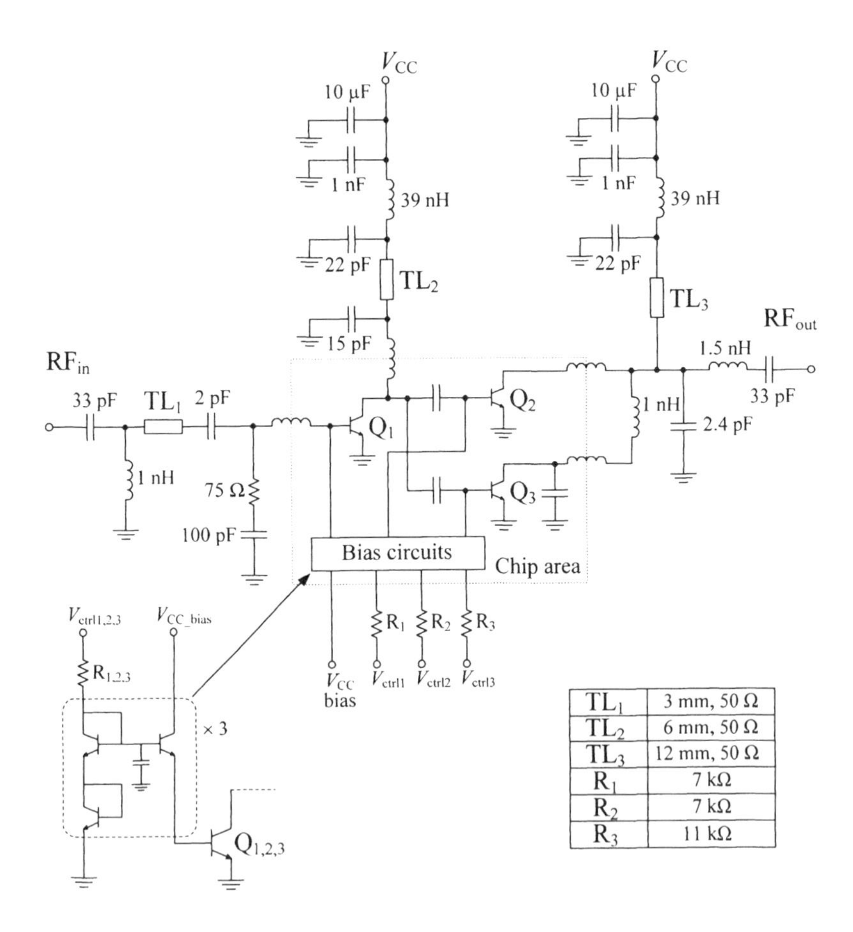

TL_1	3 mm, 50 Ω
TL_2	6 mm, 50 Ω
TL_3	12 mm, 50 Ω
R_1	7 kΩ
R_2	7 kΩ
R_3	11 kΩ

Figure 6.21. Circuit schematic of the 1.9-GHz SiGe HBT CDMA power amplifier. (Emitter areas: Q_1 = 480 μm^2, Q_2 = 1920 μm^2, and Q_3 = 480 μm^2.)

The input match network is designed to provide an optimum match of (4.0 – j12.6) Ω for the first stage. Referring to the schematic in Figure 6.21, the 75-Ω resistor and the 100-pF capacitor connected in series at the input serve as a decoupling network to reduce low-frequency gain and prevent low-frequency parasitic oscillation. The output match network is designed to create an optimum match impedance of (9.8 – j12.2) Ω for the large transistor, and an additional L section (consisting of the 1-nH inductor plus bond wire and the on-chip capacitor) is inserted to match the small transistor, which requires an optimum impedance of (15.4 + j14.8) Ω at 18 dBm output power. Interstage matching is designed using an L-section topology, where the inductor is realized by the bond wire, which also serves as the DC bias path. The on-chip bias circuits are designed using two diode-connected transistors to provide V_{BE} of the power transistor and the transistor that supplies base current to the power transistor. This topology is suitable for Class AB bias, where the base current changes over a wide range and a low-impedance bias circuit is needed.

The SiGe HBT power amplifier is tested at 1.9 GHz for the CDMA/PCS IS-95 standard. The design achieves 1-dB output compression at 27 dBm, with a maximum PAE of 27% in the high-power mode. The power amplifier is switched to low-power mode when output power is lower than 18 dBm. In the low-power mode, the maximum PAE is 15%, which is about twice the PAE of a design that has a single transistor in the output stage (see Figure 6.24). The power amplifier provides a power gain of 20 dB in the high-power mode and 18 dB in the low-power mode, as shown in Figure 6.25. The linearity of the power amplifier is characterized by the adjacent- and alternate-channel power leakage ratios (ACPR1 and ACPR2). Here, the frequency offsets from the carrier are 1.25 MHz for the adjacent channel and 2.25 MHz for the alternate channel. The IS-95 specification limits these levels to be lower than 44 dBc and 53 dBc for ACPR1 and ACPR2, respectively. The measurement results, plotted in Figure 6.26, show that the design meets the specification up to 16 dBm output power in the low-power mode and 25 dBm output power in the high-power mode. Linearity can be improved in the high-output power mode by redesigning the circuit with a larger device size in the second stage. However, the current design demonstrates that using a smaller transistor in the low-power region improves PAE while meeting the linearity requirements.

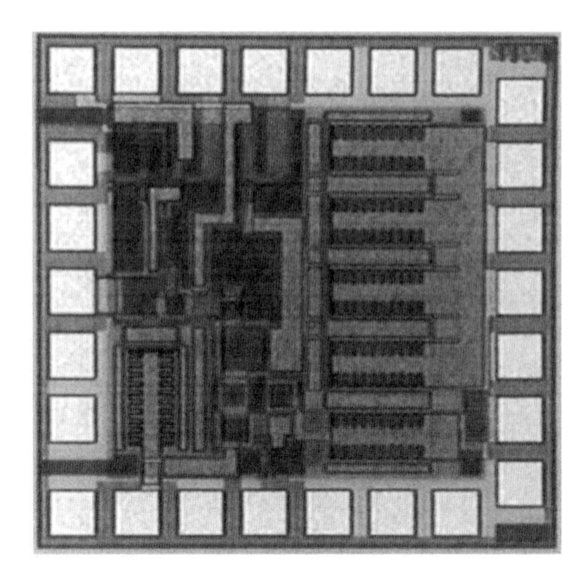

Figure 6.22. Chip photograph of the 1.9-GHz SiGe HBT CDMA power amplifier.

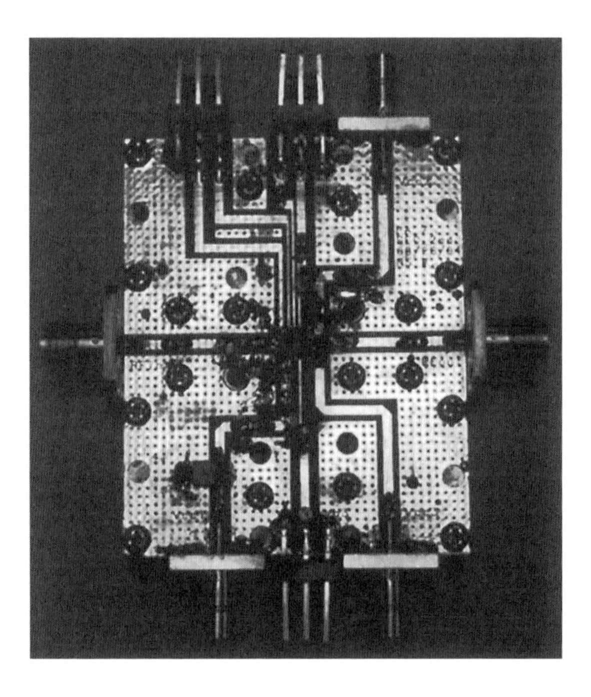

Figure 6.23. The SiGe HBT CDMA power amplifier mounted on a PCB.

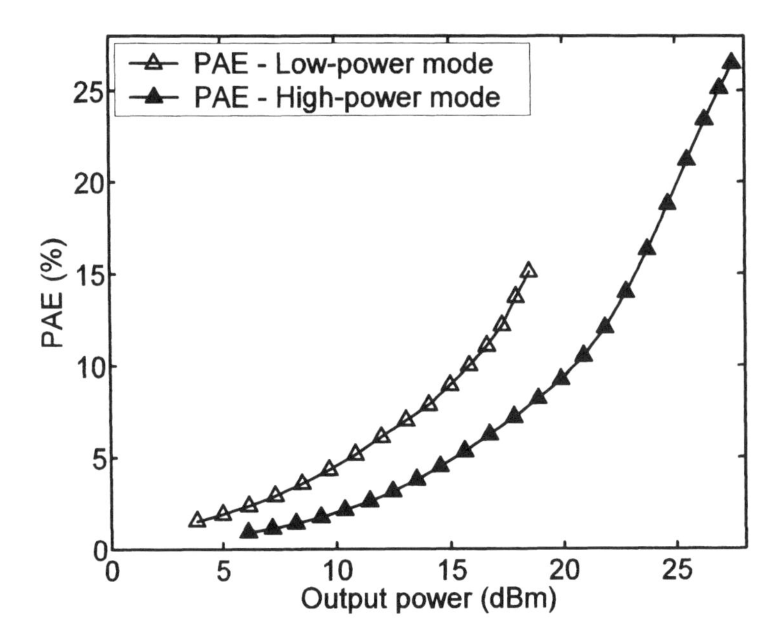

Figure 6.24. Power-added efficiency of the power amplifier at 1.9 GHz.

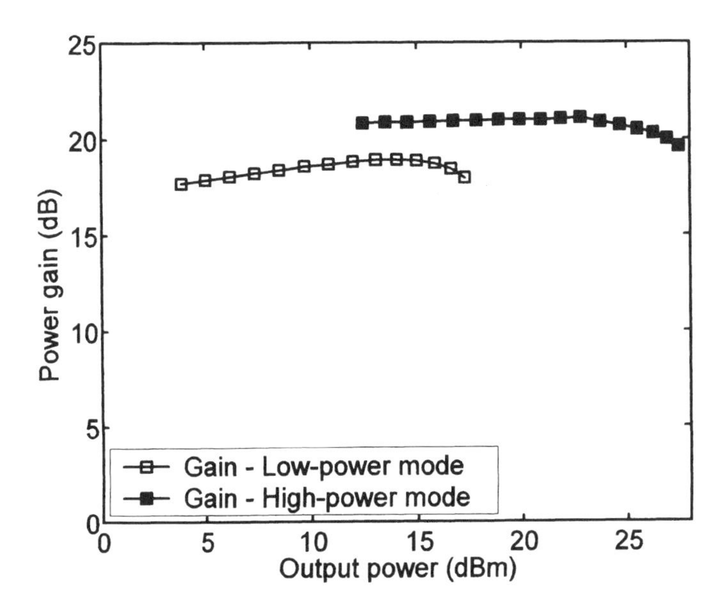

Figure 6.25. Power gain of the SiGe HBT power amplifier at 1.9 GHz.

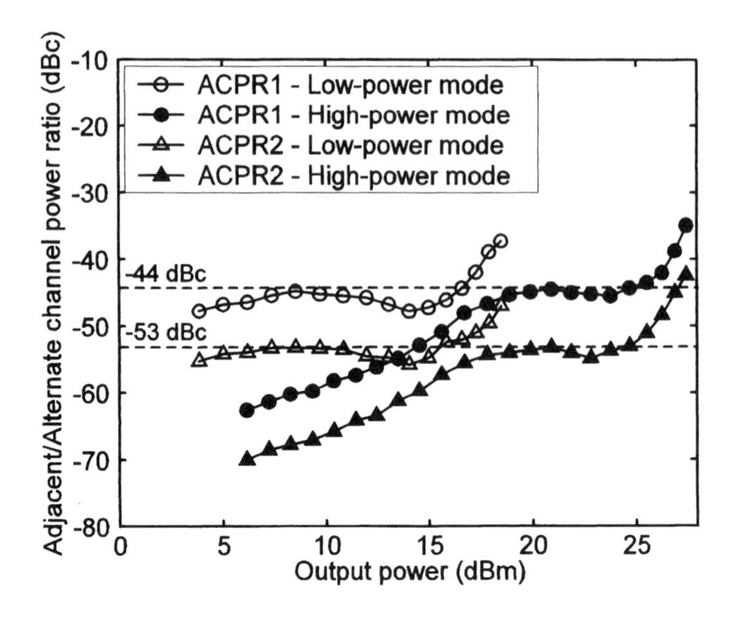

Figure 6.26. Adjacent- and alternate-channel power leakage ratios of the SiGe HBT power amplifier tested for the IS-95 standard at 1.9 GHz.

6.3.4 Nonlinear Term Cancellation for Linearity Improvement

Section 6.3.2 illustrated the use of the device periphery adjustment technique to improve the efficiency of a MOSFET power amplifier. In conjunction with this efficiency improvement, the FET power amplifier can be linearized by the cancellation of nonlinear terms. If the biases of parallel devices are carefully optimized, the nonlinear products may be designed to have opposite polarities, which results in the reduction of output intermodulation distortion (IMD). This can be explained using the Taylor series expansion of the drain current, given by

$$\begin{aligned} i_{\mathrm{d}} &= \left.\frac{dI_{\mathrm{d}}}{dV_{\mathrm{g}}}\right|_{V=V_{\mathrm{g},0}} v_{\mathrm{g}} + \frac{1}{2}\left.\frac{d^2 I_{\mathrm{d}}}{dV_{\mathrm{g}}^2}\right|_{V=V_{\mathrm{g},0}} v_{\mathrm{g}}^2 + \frac{1}{6}\left.\frac{d^3 I_{\mathrm{d}}}{dV_{\mathrm{g}}^3}\right|_{V=V_{\mathrm{g},0}} v_{\mathrm{g}}^3 + \dots \\ &= g_1 v_{\mathrm{g}} + g_2 v_{\mathrm{g}}^2 + g_3 v_{\mathrm{g}}^3 + \dots \end{aligned} \tag{6.13}$$

At different gate bias levels, g_3 may have opposite polarities: positive below threshold voltage (V_{th}) and negative above it (see Figure 6.27). Therefore, by offsetting the gate bias voltage of the output-stage transistors, their third-order intermodulation (IM3) products can be cancelled. This cancellation, however, must occur with minimal phase difference between two signal paths.

The adjustment of the gate bias voltage of the smaller transistor of the output stage (M_3 in Figure 6.17) results in an improvement in IM3 as shown in Figure 6.28. It is found experimentally that a V_{G3} of 0.55 V provides the most improvement in IM3 in the high-power mode --- about 8 dB improvement compared to the normal design, where the smaller transistor is biased in Class AB (with a V_{G3} of 0.90 V). These results are obtained from two-tone measurements with the two carrier signals at 1.88 and 1.8801 GHz (100 kHz signal separation). The output spectra from the two-tone measurements with V_{G3} = 0.9 V and V_{G3} = 0.55 V are shown in Figures 6.29 and 6.30, respectively. From these figures, we observe an IM3 improvement of 8 dB (from –25.17 dBc to –33.17 dBc) at an output power level of 19.5 dBm. Thus, the use of innovative design techniques makes it possible to achieve high PAE and linearity simultaneously in MOSFET power amplifier ICs.

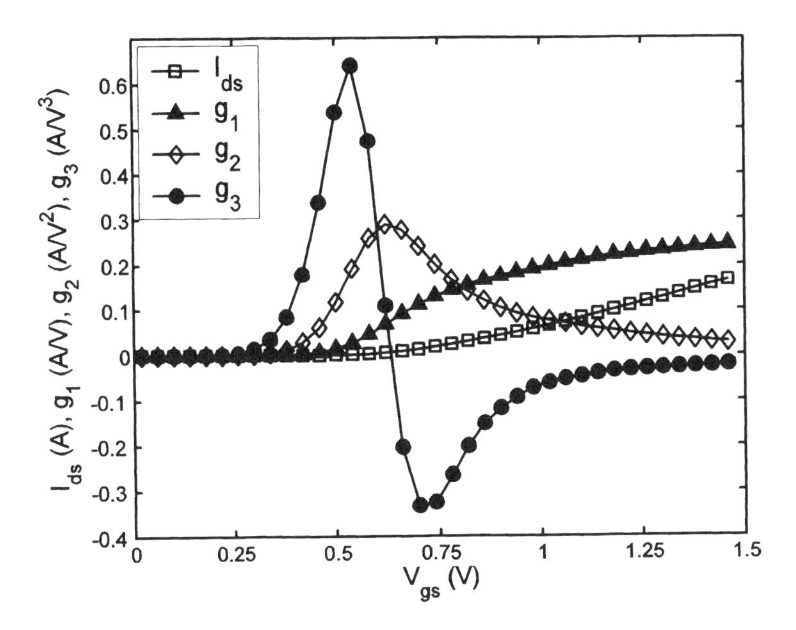

Figure 6.27. Small-signal components of the drain current (i_d) of N-MOSFET device, including the transconductance (g_1) and higher-order terms (g_2 and g_3).

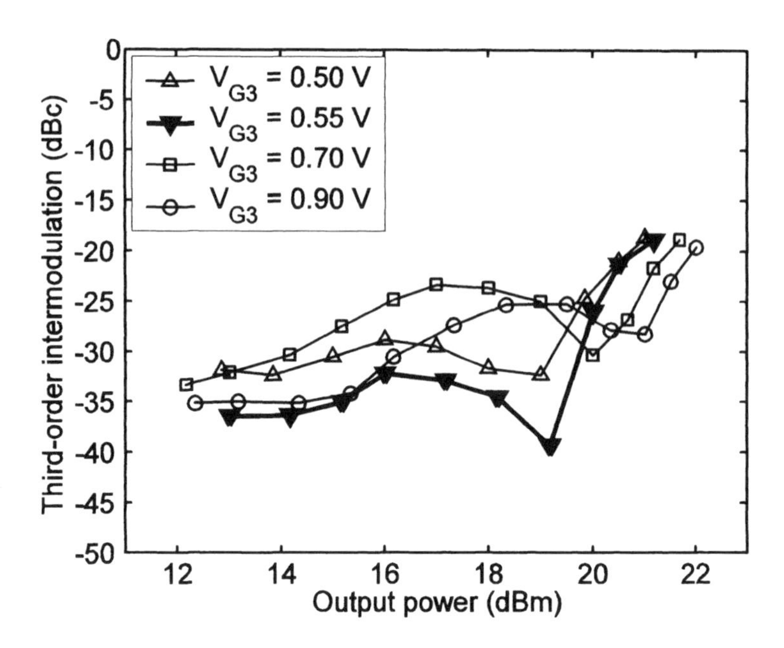

Figure 6.28. Third-order intermodulation distortion levels of the CMOS power amplifier with two-tone input signals at 1.88 and 1.8801 GHz, in the high-power mode. The gate bias voltage of M_3 is varied from 0.50 to 0.90 V.

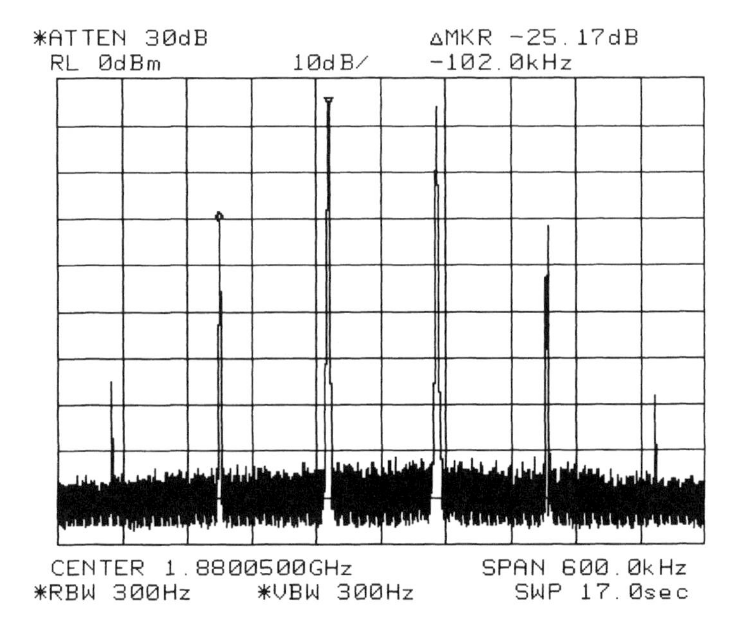

Figure 6.29. Output power spectrum at 19.5 dBm, showing carrier signals, and third- and fifth-order intermodulation products in the high-power mode, for V_{G3} = 0.9 V.

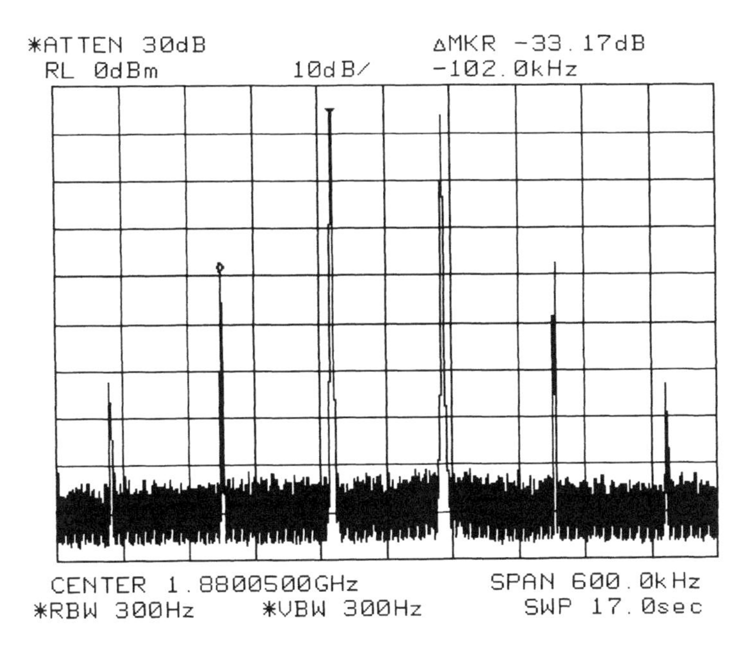

Figure 6.30. Output power spectrum at 19.5 dBm, showing carrier signals, and third- and fifth-order intermodulation products in the high-power mode, for V_{G3} = 0.55 V. An 8 dB reduction in IM3 is observed compared to the case where V_{G3} = 0.9 V.

REFERENCES

1. K. Nellis and P. J. Zampardi, A comparison of linear handset power amplifiers in different bipolar technologies, *IEEE J. Solid-State Circuits*, **39**: 1746–1754 (Oct. 2004).
2. I. Aoki, S. D. Kee, D. B. Rutledge, and A. Hajimiri, Fully integrated CMOS power amplifier design using the distributed active-transformer architechture, *IEEE J. Solid-State Circuits*, **37**: 371–383 (March 2002).
3. M. Bopp, M. Alles, M. Arens, D. Eichel, S. Gerlach, R. Gotzfried, F. Gruson, M. Kocks, G. Kimmer, R. Reimann, B. Roos, M. Siegle, and J. Zieschang, A DECT transceiver chip set using SiGe technology, *IEEE 1999 ISSCC Digest*, 1999, pp. 68–69.
4. R. Gotzfried, F. Beisswanger, S. Gerlach, A. Schuppen, H. Dietrich, U. Seiler, K.-H. Bach, and J. Albers, RFIC's for mobile communication systems using SiGe bipolar technology, *IEEE Trans. Microwave Theory Tech.*, **46**(5): 661–668 (May 1998).
5. W. Bischof, M. Alles, S. Gerlach, A. Kruck, A. Schuppen, J. Sinderhauf, H.-J. Wassener, SiGe power amplifier in flipchip and packaged technology, *IEEE 2001 RFIC-S Digest*, 2001, pp. 35–38.

6. P.-D Tseng, L. Zhang, G-B. Gao, and M. F. Chang, A 3-V monolithic SiGe HBT power amplifier for dual-mode (CDMA/AMS) cellular handset applications, *IEEE J. Solid-State Circuits*, **35**(9): 1338–1344 (Sept. 2000).
7. A. Raghavan, D. Heo, M. Maeng, A. Sutono, K. Lim, and J. Laskar, A 2.4 GHz high-efficiency SiGe HBT power amplifier with high-Q LTCC harmonic suppression filter, *2002 IEEE Int. Microwave Symp. (MTT-S) Digest*, 2002, pp. 1019–1022.
8. D. M. Pozar, *Microwave Engineering*, Addison-Wesley, Reading, MA; 1990.
9. G. L. Matthaei, L. Young, and E. M. T. Jones, *Microwave Filters, Impedance Matching Networks and Coupling Structures*, Mc-Graw Hill, New York; 1980.
10. M. Miyazaky, H. Asao, and O. Ishida, A broad band dielectric diplexer using a snake strip-line, *1991 IEEE Int. Microwave Symp. (MTT-S) Digest*, June 1991, pp. 552–554.
11. *The ARRL Handbook*, The American Radio Relay League, Newington, CT; 1992.
12. T. H. Lee, *The Design of CMOS Radio-frequency Integrated Circuits*, Cambridge University Press, Cambridge, UK; 1998.
13. F. Bergmann and D. Gerstner, Some new aspects of thermal instability of the current distribution in power transistors, *IEEE Trans. Electron Devices*, **13**(8): 630–634 (Aug. 1996).
14. R. H. Winkler, Thermal properties of high-power transistors, *IEEE Trans. Electron Devices*, **14**(5): 260–263 (May 1967).
15. P. Arnold and D. S. Zoroclu, A quantitative study of emitter ballasting, *IEEE Trans. Electron Devices*, **21**(7): 385–391 (July 1974).
16. R. T. Dennison and K. M. Walter, Local thermal effects in high performance bipolar devices/circuits, *Proc. IEEE Bipolar/BiCMOS Circuits Technology Meeting*, 1989, pp. 164–167.
17. F. Morin and J. Maita, Electrical properties of silicon containing arsenic and boron, *Phys. Rev.*, **96**: 28–35 (Oct. 1954).
18. J.-W. Sheen, A compact semi-lumped low-pass filter for harmonics and spurious suppression, *IEEE Microwave Guided Wave Lett.*, **10**(3): 92–93 (March 2000).
19. J. Staudinger, Applying switched gain stage concepts to improve efficiency and linearity for mobile CDMA power amplification, *Microwave J.*, **43**: 152–162 (Sept. 2000).
20. J. Staudinger, Stage bypassing in multi-stage PAs, paper presented at *IEEE MTT-S Symp. Workshop on Efficiency and Linearity Enhancement Methods for Portable RF/MW Power Amplifiers*, June 2000.
21. J. H. Kim, Y. S. Noh, Y. S. Kim, S. G. Kim, and C. S. Park, An MMIC smart power amplifier of 21% PAE at 16 dBm power level for W-CDMA mobile communication terminals, *Proc. 2002 IEEE GaAs IC Symp.*, Oct. 2002, pp. 181–184.
22. J. H. Kim, K. Y. Kim, Y. H. Choi, and C. S. Park, A power efficient W-CDMA smart power amplifier with emitter area adjusted for output power levels, *IEEE MTT-S Int. Microwave Symp. Digest*, 2004, pp. 1165–1168.
23. H.-M. Park, S.-H. Cheon, J.-W. Park, and S. Hong, Demonstration of on-chip appended power amplifier for improved efficiency at low power region, *2003 IEEE MTT-S Int. Microwave Symp. Digest*, June 2003, vol. 2, pp. 691–694.
24. D. R. Webster, G. Ataei, and D. G. Haigh, Low-distortion MMIC power amplifier using a new form of derivative superposition, *IEEE Trans. Microwave Theory Tech.*, **49**: 328–332 (Feb. 2001).
25. N. b. de Carvalho and J. C. Pedro, Large- and small-signal IMD behavior of microwave power amplifiers, *IEEE Trans. Microwave Theory Tech.*, **47**: 2364–2374 (Dec. 1999).

26. H.-M.Park and S. Hong, Characterization and modeling of intermodulation distortion asymmetry in HBT using large-signal model, *IEEE MTT-S Int. Microwave Symp. Digest*, 2003, pp. 773–776.

7

EFFICIENCY ENHANCEMENT OF RF POWER AMPLIFIERS

7.1 INTRODUCTION

Modern wireless communication systems use sophisticated modulation schemes to achieve high spectral efficiency. A trend that has emerged in today's wireless systems is the use of modulation schemes with significant amplitude modulation. In such systems, the power amplifier has to be highly linear, to achieve a high bit error rate, and limit spectral regrowth to acceptable levels. Linearity, in the context of amplitude modulated signals, means that the gain of the power amplifier must remain constant irrespective of output power level. Most wireless systems do not transmit at constant output power. For instance, in a mobile system, over time, output power varies to accommodate the varying distance between the base station and the mobile unit, and also to combat multipath and shadow fading. Power control techniques dynamically adjust the output power level of the mobile unit to increase battery lifetime. In fact, the output power level is far below the maximum most of the time. The peak-to-average power ratio (PAPR) of the spectrally efficient modulation formats in use today is large; it is about 4–9 dB in CDMA 2000 systems, and can be of the order of 10 dB in IEEE 802.11a/g wireless LAN systems.

Biasing the power amplifier in Class A or Class AB is the simplest way to achieve high linearity. However, these modes of operation exhibit low power-added efficiency (PAE), especially at power levels below the maximum output power. Efficiency of the power amplifier is important, particularly in mobile systems, since it directly affects talk time and battery lifetime. The large PAPR of the modulation schemes used means that, in order to maximize its overall operating efficiency, such a power amplifier will either have to operate in compression near maximum output power, to increase efficiency at backed-off output power, or operate at very low efficiency at low power to improve linearity at high power. The former may result in an unacceptable level of distortion and the latter, in impracticably low efficiency. Thus, techniques to enhance the efficiency of linear power amplifiers assume immense significance in modern wireless systems.

Efficiency enhancement techniques for RF amplifiers have been around since the 1930s and the era of the vacuum-tube amplifiers. In the case of high-power tube amplifiers, low efficiency at backed-off output power leads to excessive heat dissipation. This is because the DC power consumption is fixed in, say, Class A operation, while the output power decreases. The consequences are increased maintenance cost for cooling systems and electrical usage. This problem is relevant even today, in the high-power, solid-state amplifiers used in base stations. Thus, efficiency enhancement techniques also benefit such applications.

To date, several efficiency enhancement techniques have been developed with different approaches and varying levels of complexity. Some techniques may be more suitable than others for a given application, and sometimes they can be combined for improved results. The commonly used efficiency enhancement techniques are envelope elimination and restoration (EER), bias adaptation, the Doherty amplifier technique, and Chireix's outphasing amplifier technique. A brief discussion of these techniques, including their benefits and drawbacks follows.

7.2 EFFICIENCY ENHANCEMENT TECHNIQUES

7.2.1 Envelope Elimination and Restoration

The envelope elimination and restoration (EER) [1] technique uses a high-efficiency nonlinear amplifier as the main power amplifier to achieve the goal of high efficiency. The architecture, shown in Figure 7.1, is such that high linearity is achieved at the power amplifier subsystem level. The input signal is passed through a limiter before amplifying with the high-efficiency nonlinear power amplifier. The DC supply of the nonlinear power amplifier is modulated with the envelope of the input signal. Thus, as the name indicates, envelope information is eliminated using the limiter, and then restored by the modulated DC supply. The limiter is used to minimize AM-PM distortion, which may occur if the input to the nonlinear amplifier is too high. The output signal has amplitude proportional to the modulated DC supply voltage with the phase characteristic of the input signal.

There are many potential problems in realizing this technique: (1) phase and gain mismatch between the two paths must be minimized, which is difficult to achieve from different circuits operating at different frequencies; (2) the DC controller that generates control current and voltage for the power amplifier may not operate at 100% efficiency and has limited bandwidth, which may not be adequate for multichannel signals; and (3) the limiter circuit, and the nonlinear capacitance of the power amplifier device operating under large-signal conditions, may introduce undesirable AM-PM distortion.

7.2.2 Bias Adaptation

The bias adaptation scheme, illustrated in Figure 7.2, is similar to the EER scheme. However, the limiter circuit is unnecessary. Instead of using a nonlinear amplifier, the power transistor is operated in a linear condition, and the DC supply voltage is modulated with the envelope of the input signal. The difference between this technique and EER is that the RF input signal to the power stage contains both amplitude and phase information. The supply voltage control has more flexibility than the EER technique because it does not have to perfectly match the input envelope, which relaxes the design constraints; the tradeoff is lower drain efficiency. Since the amplifier is biased in linear operation (Class A or AB), limitations in the performance of the power stage cause problems with this technique, for example, reduction in gain at low current. Gain reduction is usually observed when the current density of a device is lowered from the optimum value, which consequently reduces PAE. Also, the design of a highly efficient DC modulator with high output current is a challenge. Nevertheless, this technique is attractive because it is simpler and more practical than EER, and has been implemented in several applications [2,3].

7.2.3 The Doherty Amplifier Technique

The Doherty amplifier [4] uses more than one amplifier to operate at different power levels. Each amplifier is biased at a different bias condition, and designed to have different load terminations, so that the system as a whole can be optimized for multiple power levels. The conventional Doherty amplifier design uses two amplifiers to compromise between efficiency and linearity in the low-power and high power regions (see Figure 7.3)

Operation of the Doherty amplifier does not require manual switching between its component amplifiers. Its operation is actually controlled by the input power level, without any use of external control or adjustment. This makes it attractive to circuit designers. Additionally, the output termination of each amplifier is controlled by the loading from another amplifier and the transformation of impedance over a quarter-wave transmission line. The output termination is automatically adjusted to maximize efficiency over a wide range of power levels. The Doherty amplifier technique is discussed in greater detail later in this chapter.

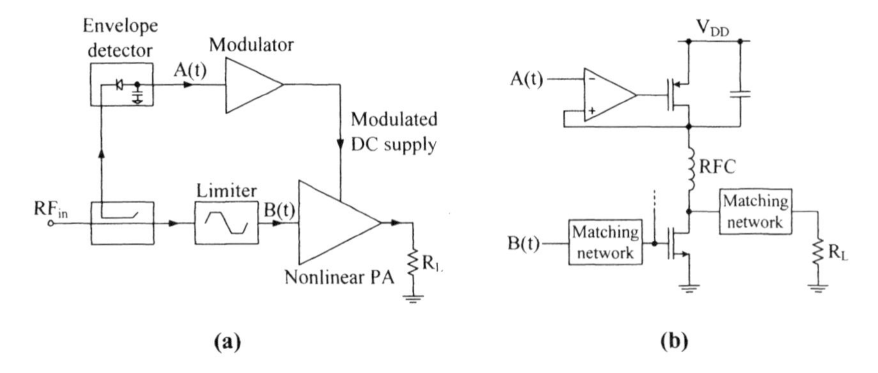

Figure 7.1. Envelope elimination and restoration: (a) system configuration; and (b) implementation of the output stage.

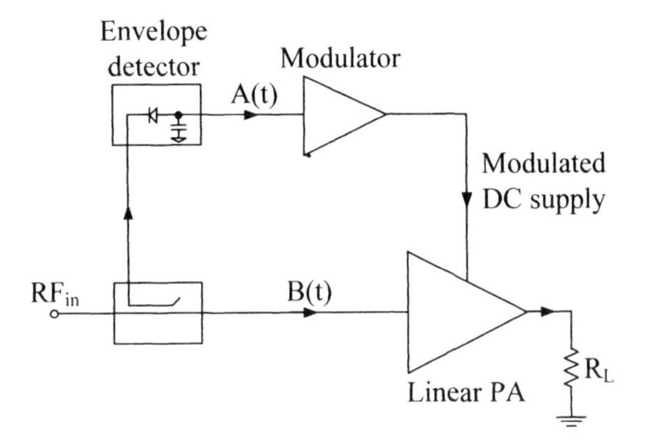

Figure 7.2. Architecture of the bias adaptation technique, used to enhance the efficiency of RF power amplifiers.

7.2.4 Chireix's Outphasing Amplifier Technique

This technique is also referred to as "linear amplification using nonlinear components (LINC)," because it uses two nonlinear amplifiers to amplify two signals with different phases (generated from the input signal), which are then combined at the output to regain the original (but amplified) amplitude- and phase-modulated signal, as illustrated in Figure 7.4. The concept of this technique was developed in 1935 by Chireix [5]. The idea originates from the trigonometric relationship,

$$\cos(A)+\cos(B)=2\cos\left(\frac{(A+B)}{2}\right)\cos\left(\frac{(A-B)}{2}\right) \tag{7.1}$$

The input signals to the two amplifiers (PA_1 and PA_2 in Figure 7.4) are

$$S_1(t)=\cos\left(\omega t+\cos^{-1}[A(t)]\right) \tag{7.2}$$

$$S_2(t)=\cos\left(\omega t-\cos^{-1}[A(t)]\right) \tag{7.3}$$

which are phase-modulated signals generated from the original input signal, $S_{in}(t)=A(t)\cos(\omega t)$, where $A(t)$ is the input amplitude. If the gain of each amplifier is G, the output signal after recombination can be written, using Equations 7.1–7.3 as

$$\begin{aligned} S_{out}(t) &= G[S_1(t)+S_2(t)] \\ &= 2GA(t)\cos(\omega t) \end{aligned} \tag{7.4}$$

Now, it is observed from Equation 7.4 that $S_{out}(t)$ is the amplitude-modulated input signal amplified by $2G$. If the initial input signal is also phase-modulated, from Equation 7.1, the phase information will be restored at the output without any change. High-efficiency nonlinear amplifiers can be used here as long as their output performances are identical.

The important component in this architecture is the AM-PM modulator used to generate $S_1(t)$ and $S_2(t)$, where the input signal amplitude is transformed into phase deviation. Also, a simple signal combiner cannot be used at the output since the signals from the two amplifiers are not synchronous in phase. However, this can be solved by reactance compensation in the load, which further improves efficiency in the backoff region, which is actually the key goal of this architecture. More details regarding load design and other practical issues are available in references [6–8].

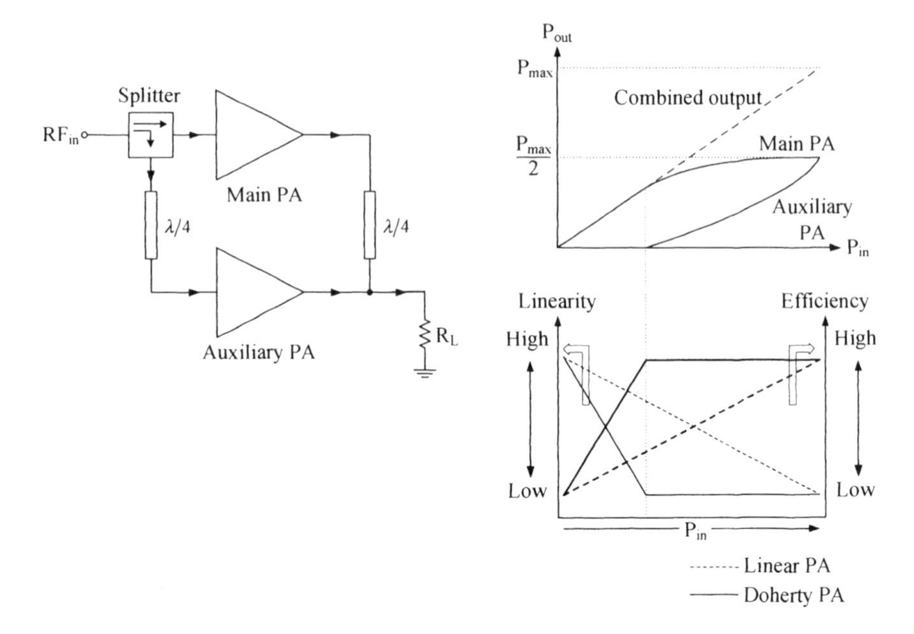

Figure 7.3. Doherty amplifier architecture and operation.

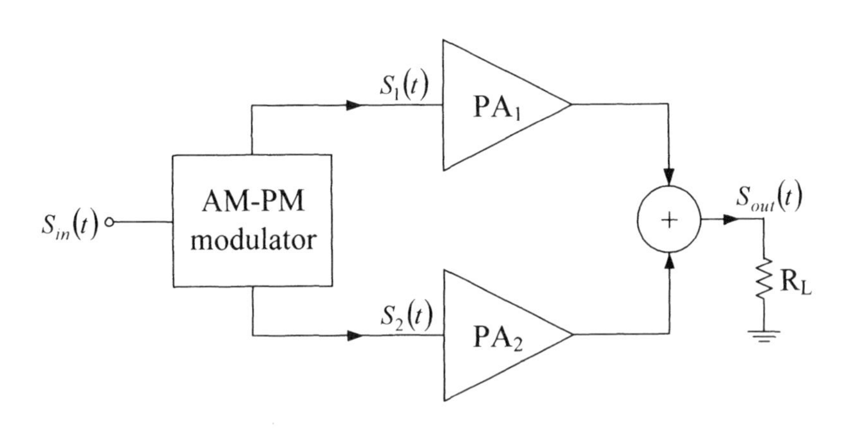

Figure 7.4. Chireix's outphasing amplifier technique.

7.3 THE CLASSICAL DOHERTY AMPLIFIER

The Doherty amplifier was first introduced in 1936 by W. H. Doherty as a method to improve the efficiency of linear vacuum-tube amplifiers in the output power backoff region. The basic idea is to provide a high-impedance termination to force the amplifier to reach an early saturation once it enters the so-called backoff region (the region just below the maximum power level where the efficiency is targeted to be maintained constant) as the input power level rises. Then the impedance termination is gradually reduced to allow the increase of output power while keeping the amplifier in saturation. This enables maintenance of maximum drain efficiency until the amplifier reaches its maximum operation level. The concept requires an adjustable impedance transformation network, which can be realized using a quarter-wave transmission line and two amplifiers (a carrier amplifier and a peak amplifier), as shown in Figure 7.5.

The operation can be mathematically explained by the impedance transformation characteristic of a quarter-wave transmission line

$$Z_{\text{in}} = \frac{Z_0^2}{Z_\text{L}} \tag{7.5}$$

where Z_{in} is the input impedance looking into the quarter-wave transmission line, Z_0 is the characteristic impedance of the transmission line, and Z_L is the load impedance. In the classical Doherty amplifier of Figure 7.5, the transmission line used to perform the Doherty operation (i.e., the automatic adjustment of the load-line) is connected to the output of the carrier amplifier. The quarter-wave transmission line at the input of the peak amplifier compensates for the phase difference between the two signal paths, so that the output signal from the two amplifiers will be added constructively at the load R_L.

In the conventional Doherty amplifier, the point where the amplifier first saturates is 6 dB below the maximum output power of the system. This means the impedance presented to the carrier amplifier in the low-power region is $4R_\text{L}$, so a characteristic impedance of $2R_\text{L}$ is required for the quarter-wave transmission line. At first, only the carrier amplifier, which is biased in Class A/AB, is in operation while the peak amplifier, which is biased below the threshold level (Class C), is not yet operating. Once the input signal reaches the predetermined level (6 dB backoff), it turns on the peak amplifier. (To do this, the bias of the peak amplifier needs to be adjusted so that it will promptly turn on at 6 dB backoff. However, this may vary from 6 dB in practice.) The impedance presented to the carrier amplifier decreases because the peak amplifier is turned on. Beyond this point, with increasing input drive, the impedance presented to the carrier amplifier constantly decreases until the system reaches the maximum power level. At this point, both carrier and peak amplifiers see a terminating impedance of $2R_\text{L}$, and the peak amplifier also reaches saturation. This implies the two amplifiers provide the same amount of power at maximum system output power and since both amplifiers are in saturation, the drain efficiency of the system will reach its maximum, as shown in Figure 7.6.

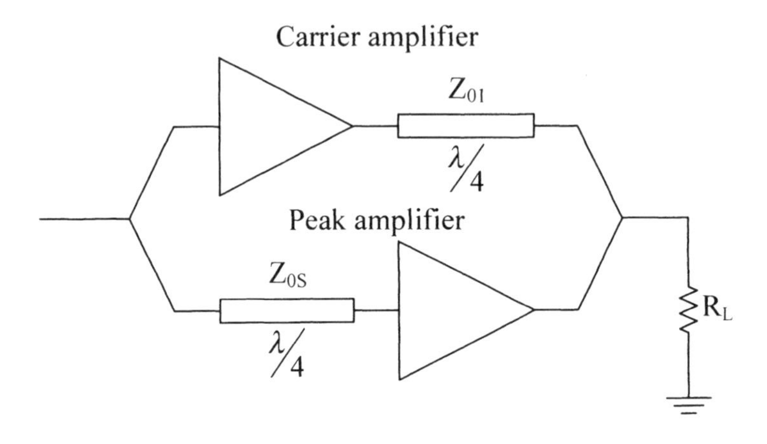

Figure 7.5. Schematic diagram of the classical Doherty amplifier.

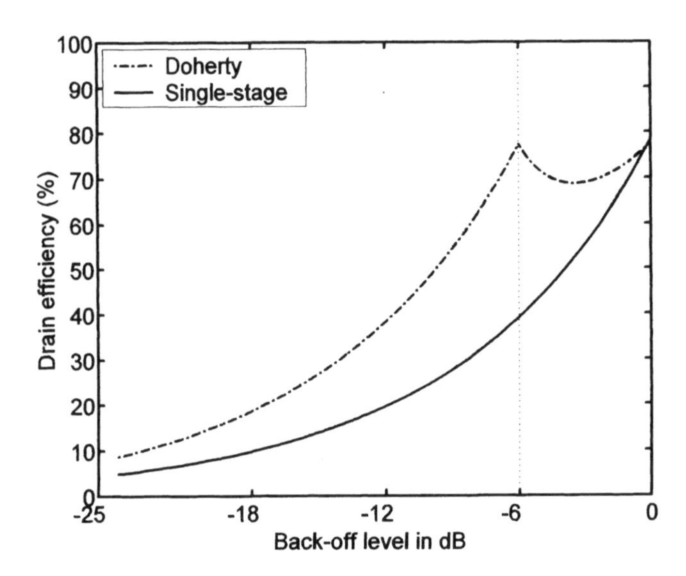

Figure 7.6. Drain efficiency of the Doherty amplifier versus output power backoff.

7.4 THE MULTISTAGE DOHERTY AMPLIFIER

The conventional Doherty amplifier offers improvement in efficiency up to 6 dB output power backoff from the maximum output power level. This is often insufficient in modern wireless communication systems, where the PAPR can be in the range of 9–12 dB. As mentioned in Section 7.1, statistically, the power usage profile shows that communication equipment usually operates at a power level far below the maximum power. Therefore, the efficiency improvement capability of the conventional Doherty amplifier still results in poor system efficiency. In this section, a method to improve the efficiency of the power amplifier at backoff levels greater than that possible with a conventional Doherty amplifier, called the *multistage Doherty amplifier*, is discussed.

The multistage Doherty amplifier, shown in Figure 7.7, uses more than one peak amplifier, with quarter-wave transmission lines to combine their output power. An active load-pulling effect is created as different amplifiers turn on at different power levels. In the design of a multistage Doherty amplifier, it is important to find the characteristic impedance of the quarter-wave transmission lines required for Doherty operation. These impedance values depend on the level of backoff and can be calculated using the following set of equations

$$Z_{0i} = R_L \prod_{j=1}^{i} \gamma_j \tag{7.6}$$

$$\prod_{j=k}^{(i+k)/2} \gamma_{2j-k} = 10^{B_i/20} \tag{7.7}$$

where $i = 1, 2, 3, \ldots, (N-1)$; $k = 1$ (for odd i) or 2 (for even i); N is the total number of amplifier stages; and B_i is the backoff level (positive value in dB) from the maximum output power of the system at which the efficiency will peak. The maximum level of backoff B_{N-1} is set by the carrier amplifier. The number of efficiency peaking points is directly proportional to the number of amplifier stages used in the design.

7.4.1 Principle of Operation

In this section, we analyze the fundamental operation of the multistage Doherty amplifier architecture. To simplify the analysis, the equivalent circuit of the three-stage Doherty amplifier with an ideal current source to represent each amplifier, as shown in Figure 7.8, is used. The design equations for an N-stage Doherty amplifier can easily be generalized from this analysis.

From Equations 7.6 and 7.7, the characteristic impedances of the two output quarter-wave transmission lines are

$$Z_{01} = \gamma_1 R_L \tag{7.8}$$

$$Z_{02} = \gamma_1 \gamma_2 R_L \tag{7.9}$$

where

$$\gamma_1 = 10^{B1/20} \tag{7.10}$$

$$\gamma_2 = 10^{B2/20} \tag{7.11}$$

The phase of the peak amplifier output currents, i_{P1} and i_{P2}, must lag that of the carrier amplifier output current, i_{C1} by 90° and 180°, respectively, for proper Doherty amplifier operation. This is achieved by inserting additional transmission lines at the inputs of the peak amplifiers. By doing so, the signal from each amplifier will be added constructively at the output load.

The operation of a three-stage Doherty amplifier can be separated into three regions: low-power operation, where only the carrier amplifier is turned on; medium-power operation, where the carrier amplifier and peak amplifier 1 are both turned on; and high-power operation where all the power amplifiers are turned on. At low-power operation (Figure 7.8a), where all the peak amplifiers are in the OFF state and appear as open circuits, the carrier amplifier sees an impedance given by

$$R_{C1}^l = \frac{(Z_{02})^2}{R_{C2}^l} = \gamma_2^2 R_L \tag{7.12}$$

where

$$R_{C2}^l = \frac{(Z_{01})^2}{R_L} = \gamma_1^2 R_L \tag{7.13}$$

Assuming that the system supply voltage is V_{DD}, the maximum output power at the load R_L, when the carrier amplifier is in saturation, is

$$P_L = \frac{V_{DD}^2}{2\gamma_2^2 R_L} \tag{7.14}$$

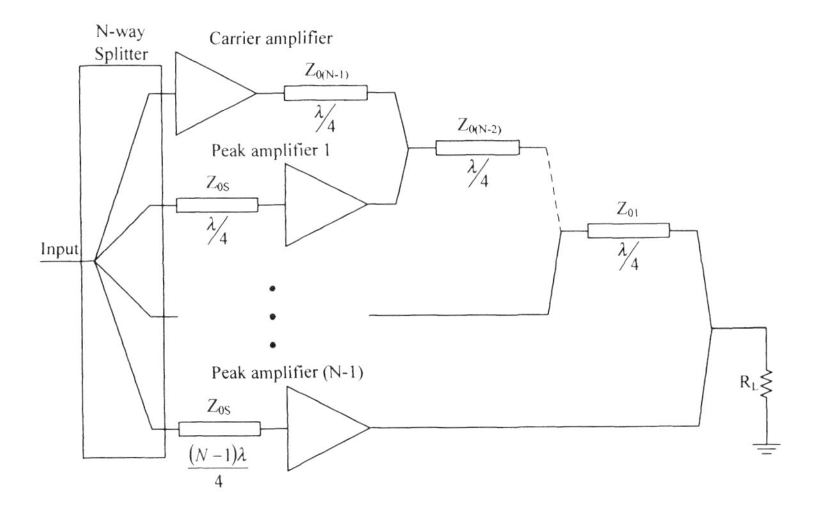

Figure 7.7. The multistage Doherty amplifier.

For the case of the three-stage Doherty amplifier, with peak efficiency at 6 and 12 dB backoff ($B_1 = 6$ and $B_2 = 12$), the values of γ_1 and γ_2 are 2 and 4, respectively. At this point, the maximum power from the carrier amplifier is one-sixteenth of the maximum possible system power. Since the other amplifiers are still turned off and the carrier amplifier is in saturation, the entire output power is delivered only from this amplifier, and therefore the overall efficiency is equal to the maximum efficiency of the carrier amplifier.

In the medium-power operation region (Figure 7.8b), the impedances presented at the outputs of the amplifiers can be analyzed using power conservation analysis. From this, the output power delivered to the load R_L (in Figure 7.8b) is equal to the combination of the output power delivered from the carrier amplifier and the peak amplifier 1, which can be expressed as

$$\frac{\left(v_L^m\right)^2}{R_L} = \frac{\left(v_{P1}^m\right)^2}{R_{C2}^m} = \frac{\left(v_{C1}^m\right)^2}{R_{C1}^m} + \frac{\left(v_{P1}^m\right)^2}{R_{P1}^m} \tag{7.15}$$

or

$$\frac{a_1^2}{R_{C2}^m} = \frac{1}{R_{C1}^m} + \frac{a_1^2}{R_{P1}^m} \tag{7.16}$$

where $a_1 = v_{P1}^m / v_{C1}^m$; and R_{C2}^m is the parallel combination of $R_{C1}^{m'}$ and R_{P1}^m, and is equal to $Z_{01}{}^2 / R_L$. Since $R_{C1}^{m'}$ is equal to $Z_{02}{}^2 / R_{C1}^m$, the expression for R_{C2}^m can be written as

$$\frac{1}{R_{C2}^m} = \frac{R_{C1}^m}{Z_{02}{}^2} + \frac{1}{R_{P1}^m} \tag{7.17}$$

Using Equations 7.16 and 7.17, R_{C1}^m and R_{P1}^m can be derived as

$$R_{C1}^m = \frac{Z_{02}}{a_1} \tag{7.18}$$

$$R_{P1}^m = \frac{a_1 Z_{02} R_{C2}^m}{a_1 Z_{02} - R_{C2}^m} \tag{7.19}$$

which leads to the following expressions:

$$R_{C1}^m = \frac{\gamma_1 \gamma_2}{a_1} R_L \tag{7.20}$$

$$R_{P1}^m = \frac{a_1 \gamma_1^2 \gamma_2}{a_1 \gamma_2 - \gamma_1} R_L \tag{7.21}$$

where $a_1 = v_{P1}^m / v_{C1}^m$, which has a value between γ_1 / γ_2 (when only the carrier amplifier saturates) and 1 (when both carrier and peak amplifier 1 saturate), assuming every amplifier uses the same drain bias voltage V_{DD} . For the case of the three-stage Doherty amplifier, with $\gamma_1 = 2$ and $\gamma_2 = 4$, R_{C1}^m reduces from $16R_L$ to $8R_L$ at the saturation of peak amplifier 1 because of the load-pulling effect resulting from the turn-on of peak amplifier 1. Also, peak amplifier 1 is presented with a transformed impedance R_{P1}^m of $8R_L$. Therefore, the total output power, which is delivered equally from both amplifiers, will be one-fourth of the maximum output power (i.e., 6 dB backoff) and will result in an efficiency peak (equal to the maximum efficiency) since both amplifiers are in saturation.

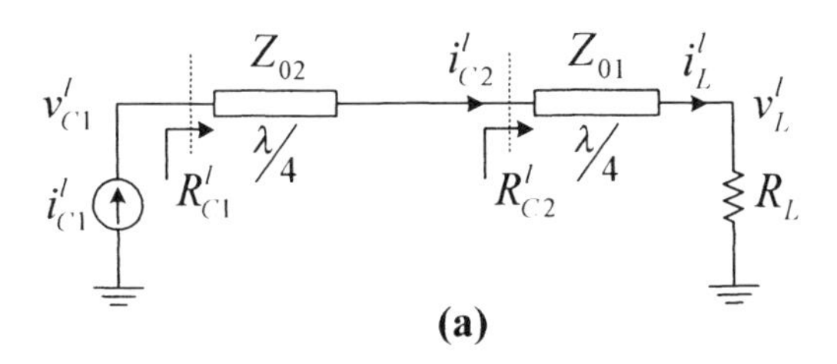

(a)

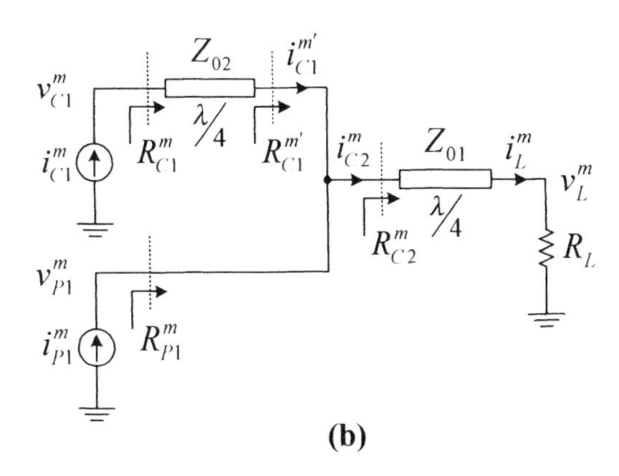

(b)

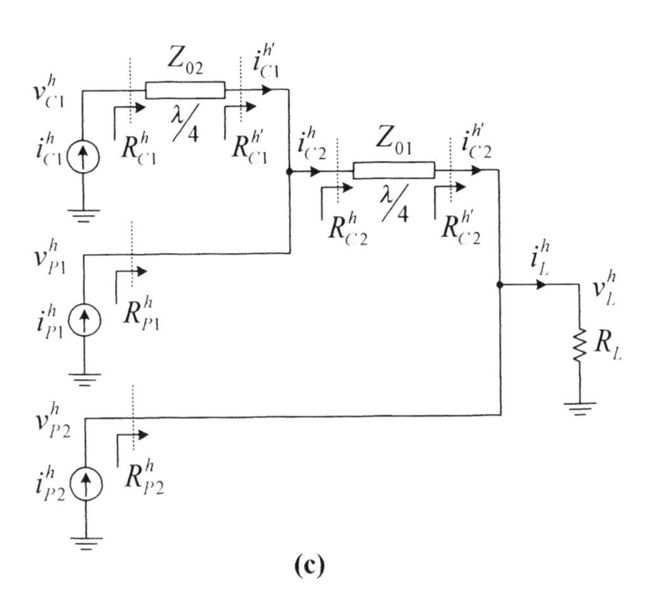

(c)

Figure 7.8. Equivalent circuit for the three-stage Doherty amplifier at (a) low-power operation, (b) medium-power operation, and (c) high-power operation.

In the high-power region (Figure 7.8c), peak amplifier 2 is in operation and produces a load-pulling effect on the other amplifiers. An analysis of the terminating impedance to the output of each amplifier can be done similarly to the medium power case, resulting in the following design equations:

$$R_{C1}^{h} = \gamma_1 \gamma_2 R_L \tag{7.22}$$

$$R_{P1}^{h} = \frac{\gamma_1 \gamma_2}{a_2 \gamma_2 - 1} R_L \tag{7.23}$$

$$R_{P2}^{h} = \frac{a_2 \gamma_1}{a_2 \gamma_1 - 1} R_L \tag{7.24}$$

where $a_2 = v_{P2}^{h} / v_{P1}^{h}$, which has a value between $1/\gamma_1$ (when peak amplifier 1 saturates) and 1 (when all amplifiers saturate). For the three-stage Doherty amplifier with $\gamma_1 = 2$ and $\gamma_2 = 4$, R_{P1}^{h} decreases from $8R_L$ to $8R_L/3$ when peak amplifier 2 has reached saturation, while the R_{C1}^{h} remains unchanged. At saturation, R_{P2}^{h} is equal to $2R_L$, which enables half the total system output power to be delivered from peak amplifier 2. The rest of the output power is delivered from the carrier amplifier and peak amplifier 1 in the ratio of 1:3.

The value of R_L should be determined by considering the amount of power delivered to the load R_L from the peak amplifier stage (N-1), since this peak amplifier will contribute more power compared to the other stages. Using Equation 7.24, peak amplifier (N-1) will be terminated with the optimum matching condition when the value of R_L is calculated by

$$\mathrm{R_L} = \frac{\gamma_1 - 1}{\gamma_1} \mathrm{R}_{\mathrm{opt\_(N-1)}} \tag{7.25}$$

$\mathrm{R}_{\mathrm{opt\_(N-1)}}$ is the optimum termination of peak amplifier (N-1).

7.4.2 Analysis of Efficiency

The efficiency of the three-stage Doherty amplifier can be calculated using power conservation analysis. For low-power operation (Figure 7.8 a), using Equation 7.12, and the relationship of the carrier amplifier DC current (I_{C1}^{l}) and voltage across the load (v_L^{l}) given by [9]

$$I_{C1}^{l} = \frac{v_L^l}{\gamma_2 R_L} \tag{7.26}$$

the ideal drain efficiency using a Class B amplifier for the carrier amplifier can be formulated as

$$\begin{aligned} \eta^{l} &= \frac{\left(v_L^l\right)^2 / 2R_L}{(2/\pi) I_{C1}^{l} V_{DD}} \\ &= \frac{\pi}{4} \frac{\gamma_2 v_L^l}{V_{DD}}, \quad 0 \le v_L^l \le \frac{V_{DD}}{\gamma_2} \end{aligned} \tag{7.27}$$

This implies a peak efficiency at $v_L^l = V_{DD}/\gamma_2$ or $P_L = 1/\gamma_2^{\ 2}$ of the maximum output power of the system ($\eta^l = 78.5\%$ for a Class B amplifier).

For medium-power operation (Figure 7.8b), DC current consumption of the carrier amplifier and peak amplifier 1 can be calculated as

$$I_{C1}^{m} = \frac{v_{P1}^{m}}{\gamma_1 \gamma_2 R_L} = \frac{V_L^m}{\gamma_2 R_L} \tag{7.28}$$

$$I_{P1}^{m} = I_{C2}^{m} - I_{C1}^{m'} = \frac{v_L^m}{\gamma_1 R_L} - \frac{V_{DD}}{\gamma_1 \gamma_2 R_L} \tag{7.29}$$

Therefore, for Class B carrier and peak amplifiers, the drain efficiency can be formulated as

$$\begin{aligned} \eta^{m} &= \frac{\left(v_L^m\right)^2 / 2R_L}{(2/\pi)\left(I_{C1}^{m} + I_{P1}^{m}\right) V_{DD}} \\ &= \frac{\pi}{4} \frac{\gamma_1 \gamma_2 \left(v_L^m / V_{DD}\right)^2}{\left(\gamma_1 + \gamma_2\right)\left(\dfrac{v_L^m}{V_{DD}}\right) - 1}, \quad \frac{V_{DD}}{\gamma_2} < v_L^m \le \frac{V_{DD}}{\gamma_1} \end{aligned} \tag{7.30}$$

From Equation 7.30, there exists another peak in efficiency at $v_L^m = V_{DD}/\gamma_1$ or $P_L = 1/\gamma_1^{\ 2}$ of the maximum output power of the system.

Similarly, for high-power operation (Figure 7.8c), the carrier and peak amplifier DC currents are derived as

$$I_{\mathrm{C1}}^{\mathrm{h}} = \frac{V_{\mathrm{DD}}}{\gamma_1 \gamma_2 R_{\mathrm{L}}} \tag{7.31}$$

$$I_{\mathrm{P1}}^{\mathrm{h}} = I_{\mathrm{C2}}^{\mathrm{h}} - I_{\mathrm{C1}}^{\mathrm{h'}} = \frac{v_{\mathrm{L}}^{\mathrm{h}}}{\gamma_1 R_{\mathrm{L}}} - \frac{V_{\mathrm{DD}}}{\gamma_1 \gamma_2 R_{\mathrm{L}}} \tag{7.32}$$

$$I_{\mathrm{P2}}^{\mathrm{h}} = I_{\mathrm{L}}^{\mathrm{h}} - I_{\mathrm{C2}}^{\mathrm{h'}} = \frac{v_{\mathrm{L}}^{\mathrm{h}}}{R_{\mathrm{L}}} - \frac{V_{\mathrm{DD}}}{\gamma_1 R_{\mathrm{L}}} \tag{7.33}$$

Finally, the drain efficiency is given by

$$\begin{aligned} \eta^{\mathrm{h}} &= \frac{\left(v_{\mathrm{L}}^{\mathrm{h}}\right)^2 / 2R_{\mathrm{L}}}{(2/\pi)\left(I_{\mathrm{C1}}^{\mathrm{h}} + I_{\mathrm{P1}}^{\mathrm{h}} + I_{\mathrm{P2}}^{\mathrm{h}}\right)V_{\mathrm{DD}}} \\ &= \frac{\pi}{4} \cdot \frac{\gamma_1 \left(v_{\mathrm{L}}^{\mathrm{h}} / V_{\mathrm{DD}}\right)^2}{(\gamma_1 + 1)\left(v_{\mathrm{L}}^{\mathrm{h}} / V_{\mathrm{DD}}\right) - 1}, \quad \frac{V_{\mathrm{DD}}}{\gamma_1} < v_{\mathrm{L}}^{\mathrm{h}} \leq V_{\mathrm{DD}} \end{aligned} \tag{7.34}$$

Equation 7.34 shows a peak in efficiency at $v_{\mathrm{L}}^{\mathrm{h}} = V_{\mathrm{DD}}$, which is at maximum system power. For the general case of the N-stage Doherty amplifier, the drain efficiency can be calculated similarly to the analysis above. The simulated drain efficiency of the multistage Doherty amplifier using Class B amplifiers is shown in Figure 7.9.

7.4.3 Practical Considerations

In the Doherty amplifier, peak amplifiers are biased below the threshold so that they will not be turned on before the input power has reached a predetermined level. This means that the operation of peak amplifiers has been forced into Class C, which is known to typically have lower gain compared to the Class A/AB operation of the carrier amplifier. Therefore, to achieve the required output power in each operation region, two factors can be controlled: the input power to each amplifier; and the size of each transistor, which is related to current gain. It is possible to provide greater input power to the peak amplifiers, which are biased in Class C, to achieve the required output power without creating AM-AM distortion. Nevertheless, the unequal division of input power will reduce the amount of output power from other transistors, thus reducing the overall gain. To avoid this, the current gain of each transistor must be adjusted to compensate for the gain reduction in the Class C amplifier with minimum alteration of the input power division. To illustrate this, the three-stage Doherty amplifier described in the last section is used in the following analysis.

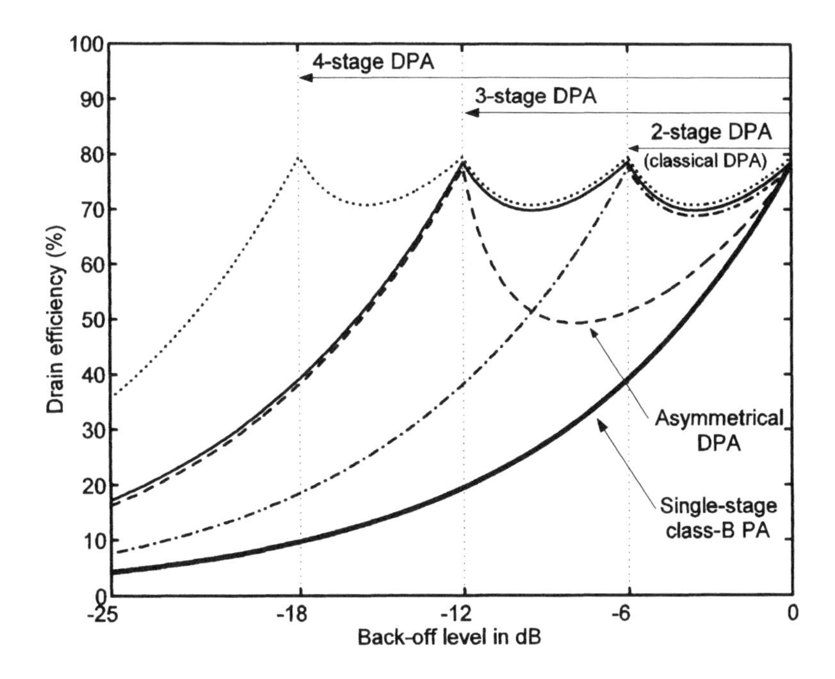

Figure 7.9. Drain efficiency of the multistage Doherty power amplifier (DPA) using Class B amplifiers ($\gamma_1 = 2$, $\gamma_2 = 4$, $\gamma_3 = 4$; $\gamma = 4$ for asymmetric DPA [10]).

To maintain the gain level up to maximum power, the periphery of the peak amplifiers needs to be increased to compensate for fundamental current reduction (assuming that a fixed bias configuration is used). The fundamental component of RF current (I_{fund}) is given by the following expression [6]:

$$I_{\text{fund}} = \frac{I_{\text{sat}}(\theta - \sin\theta)}{2\pi(1 - \cos(\theta/2))} \tag{7.35}$$

where I_{sat} is the maximum current swing for Class A operation and θ is the conduction angle.

Since I_{sat} is proportional to the device periphery, using Equations 7.31–7.33, and assuming that, in a general case, peak amplifier transistors are approaching Class B bias at peak system operation, the relative device periphery can be calculated to be

$$\text{Device periphery ratio} = 1 \;:\; \gamma_2 - 1 \;:\; \gamma_2(\gamma_1 - 1) \tag{7.36}$$

(carrier amplifier : peak amplifier 1 : peak amplifier 2)

In practice, the periphery of the transistor for peak amplifiers may need to be larger than the values calculated above as they may still be in Class C bias at maximum system operation, as shown in Figure 7.10. This can be determined experimentally depending on the extent of backoff efficiency improvement and bias point design. Nevertheless, the periphery of the carrier amplifier must be adjusted to provide sufficient output power at the system's maximum operating level and used as a reference for the peak amplifiers. The periphery of the carrier amplifier device can be calculated by

$$DP_C = \left(\prod_{j=1}^{N-1} \gamma_j \right)^{-1} \times DP_{P_L \max} \tag{7.37}$$

where DP_C is the device periphery of the carrier amplifier and $DP_{P_L \max}$ is the device periphery of a Class A biased transistor that can deliver the system's maximum power to the load R_L.

7.4.4 Measurement Results

To verify the analysis, a three-stage Doherty power amplifier with $\gamma_1 = 2$ and $\gamma_2 = 4$ was designed using GaAs FET devices and microstrip-based power combining elements on an FR-4 printed circuit board. The schematic diagram is shown in Figure 7.11, with the actual board layout shown in Figure 7.12. The three-stage Doherty power amplifier is targeted for a Class 1 WCDMA uplink standard with −33 dBc ACLR1 and −43 dBc ACLR2 linearity requirements, with 33 dBm maximum output power. The input signal is divided equally by a microstrip Wilkinson power divider network before feeding to the input matching network of each amplifier. Delay lines of 90° and 180° are inserted at the input of peak amplifiers 1 and 2, respectively. The design is tested with a single-tone signal at 1.95 GHz, the results of which are shown in Figure 7.13.

The choice of device periphery is calculated from Equation 7.36 with the predetermined backoff level improvement, which results in a ratio of 1:3:4. However, because of limitations in device size availability, a device size ratio of 1:2:4 was chosen with a W/L of the carrier amplifier device of 2400 μm/0.6 μm. The drain bias voltage for all GaAs FET devices in this design is 10 V. The gate bias voltage of the carrier amplifier is set to −1.64 V, which is above the pinchoff voltage of −2 V, for Class AB biasing. The gate bias voltages of peak amplifiers 1 and 2 are adjusted so that they are turned on at the backed-off levels of 12 and 6 dB, respectively.

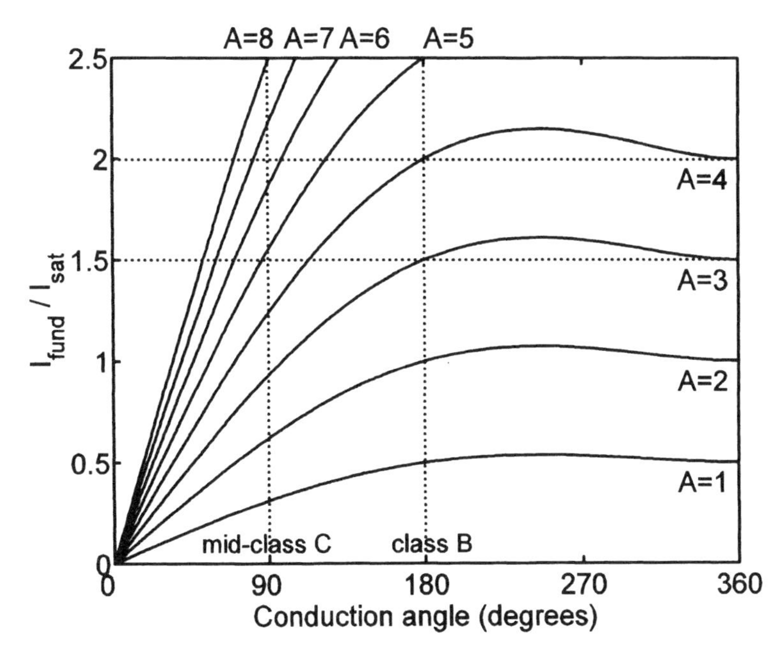

Figure 7.10. Fundamental current component as a function of conduction angle (A= relative device periphery).

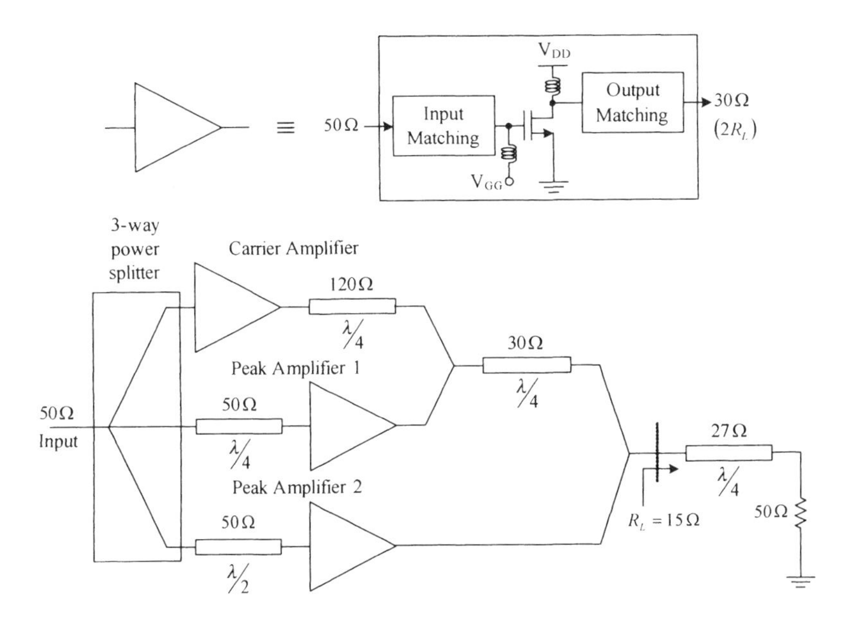

Figure 7.11. Schematic diagram of the three-stage WCDMA Doherty power amplifier.

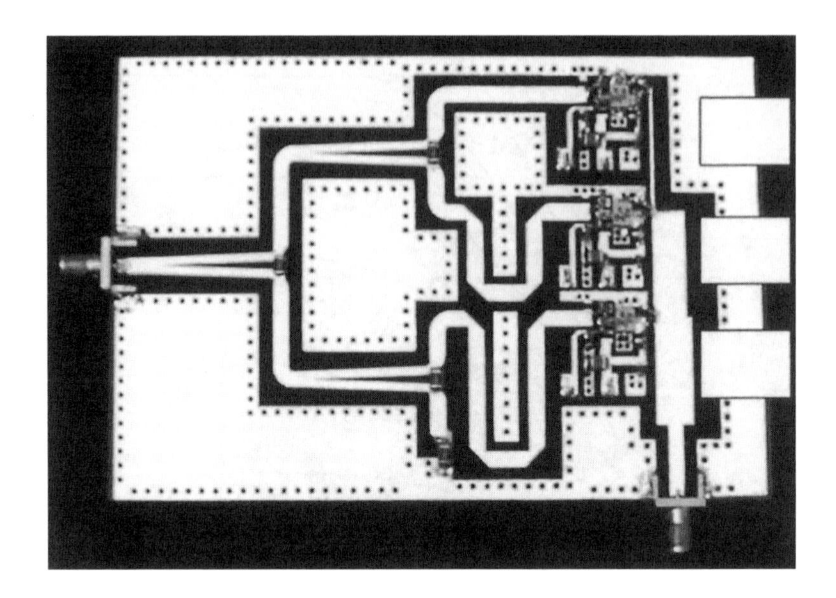

Figure 7.12. Board layout of the three-stage WCDMA Doherty power amplifier prototype using a device periphery ratio of 1:2:4. Each GaAs FET device has the same package size of 3.8 × 4.2 mm.

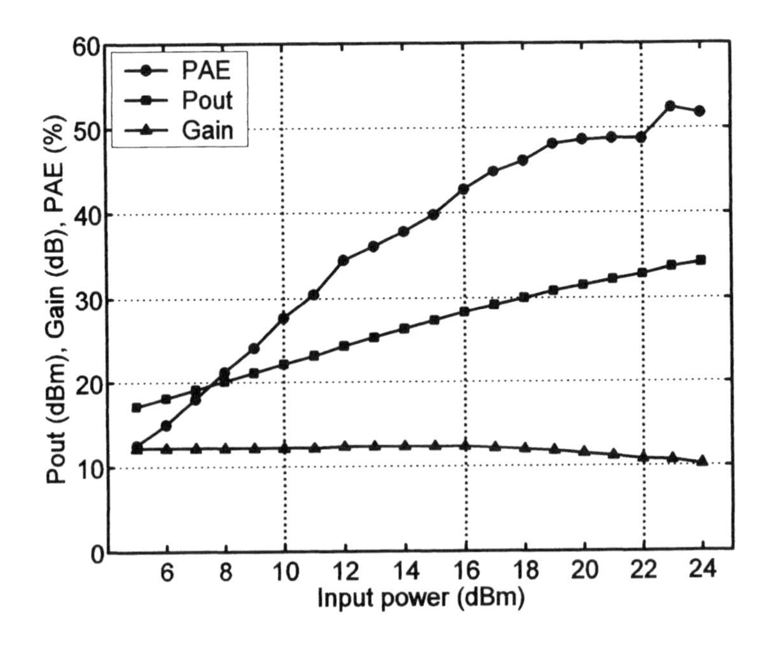

Figure 7.13. Measured output power, gain, and PAE of the three-stage WCDMA Doherty power amplifier using a device periphery ratio of 1:2:4 at 1.95 GHz.

The design achieved a linear power gain of 12.2 dB with 1-dB output compression at 33 dBm. The 33 dBm output power level is defined as the 0 dB backoff level. At that point, the power-added efficiency (PAE) was measured to be 48.5%. Also, the PAE was measured to be 42% at 6 dB backoff, and 27% at 12 dB backoff, which represents a PAE improvement of 2.5 times and 7.5 times, respectively, compared to a single-stage Class AB design. The PAE at 12 dB backoff, shown in Figure 7.14, is lower than the ideal simulation mainly because of the soft turnon characteristic of the peak amplifier devices, which contributes to more DC power consumption at low power level. Nevertheless, these results show an impressive PAE improvement in the backoff region. For comparison, a three-stage Doherty power amplifier with $\gamma_1 = 2$ and $\gamma_2 = 4$ using a device periphery ratio of 1:1:1 was designed and measured. It is seen from Figure 7.14 that the lower device periphery ratio (1:1:1) results in poorer PAE improvement at the same backoff level.

The three-stage Doherty power amplifier is tested with a real-time WCDMA 3GPP signal using a chip rate of 3.84 Mcps (megachips per second). The output signal is measured with a raised-root-cosine (RRC) filter with α of 0.22 and a bandwidth equal to the chip rate. The adjacent channel power leakage ratio (ACLR) measurement results in Figure 7.15 show that the design meets the WCDMA ACLR requirements of −33 dBc (ACLR1) and −43 dBc (ACLR2) at 5 MHz and 10 MHz offset, respectively, up to a power output of 34 dBm. The ACLR1 and ACLR2 at 33 dBm output power are measured to be −35 dBc and −47 dBc, which provide a few dB of margin over the linearity requirements. Moreover, it can be noticed that the ACLR levels of the three-stage Doherty power amplifier with the device periphery ratio of 1:2:4 are higher than that of the 1:1:1 design since each amplifier stage is operated closer to its saturation, resulting in better overall PAE. The 3 dB bandwidth is measured to be 160 MHz, dominated by the quarter-wave transformer characteristic, with only ±0.5 dB output power variation from center frequency in the WCDMA uplink frequency range (1.92–1.98 GHz), as shown in Figure 7.16. The ACLR is still within the specification in this frequency range (Figure 7.17).

Thus, the device size ratio of 1:2:4 obtained from ideal calculations results in performance sufficient to meet the stringent WCDMA requirements. However, in reality, the peak amplifier devices may not have reached Class B operation, as assumed in the ideal-case calculations. It is possible to further enhance the performance of the Doherty power amplifier by moving the bias point of the peak amplifier from Class C in the backoff region toward Class B as much as possible, by adjusting the gate voltage with increasing input drive. In this design, peak amplifier 2, which was biased in very deep Class C, is now biased manually with a dynamic bias profile, as shown in Figure 7.18.

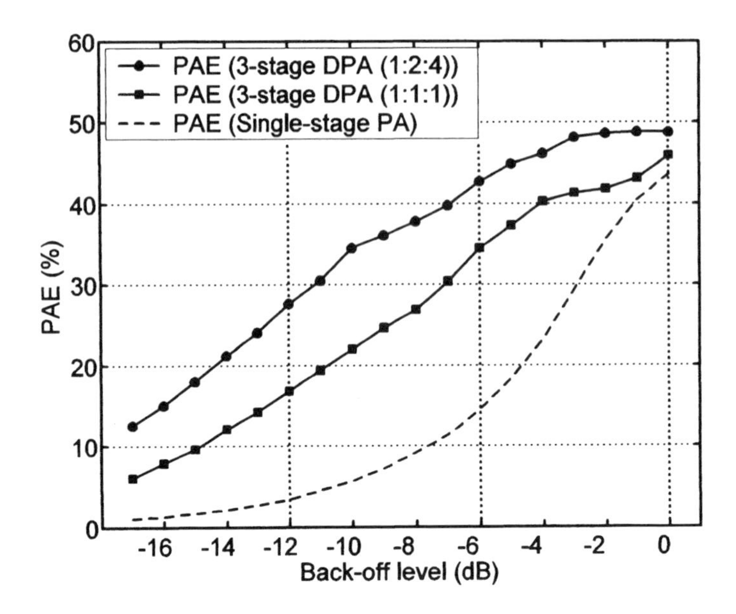

Figure 7.14. Comparison of PAE measurements of three-stage WCDMA Doherty power amplifiers using different device periphery ratios at 1.95 GHz (0 dB backoff corresponds to output power of 33 dBm).

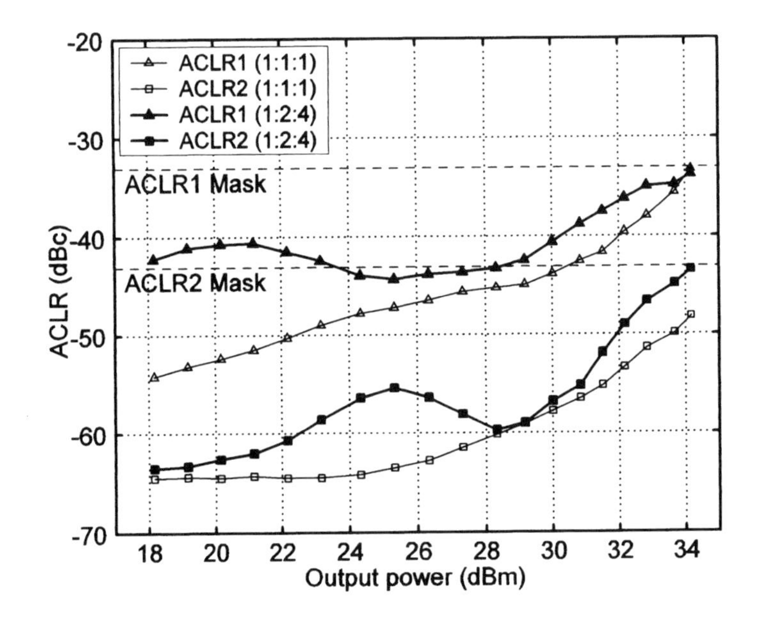

Figure 7.15. Measured ACLRs of the three-stage WCDMA Doherty power amplifiers using different device periphery ratios, at 1.95 GHz.

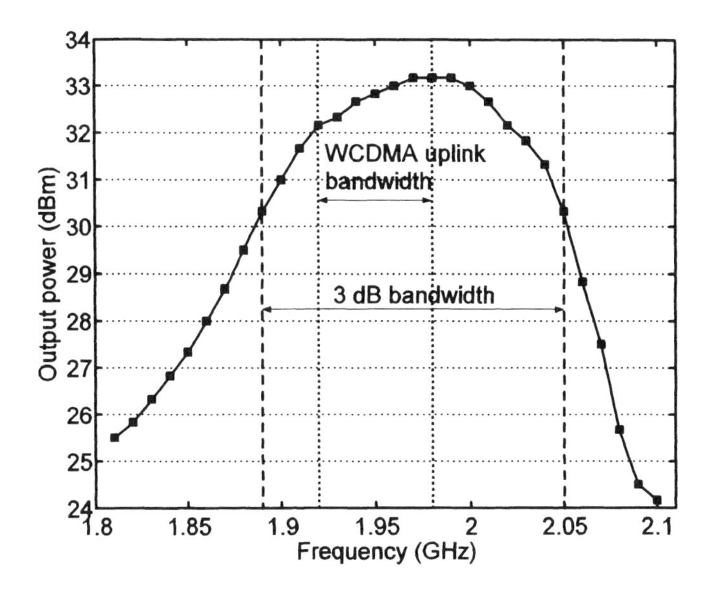

Figure 7.16. Measured output power of the three-stage WCDMA Doherty power amplifier with a device periphery ratio of 1:2:4 versus frequency, at the maximum output power level (33 dBm at 1.95 GHz).

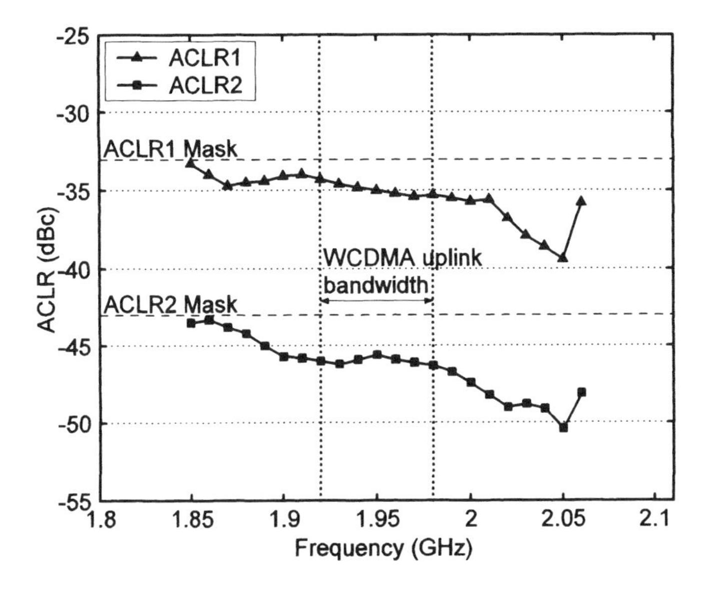

Figure 7.17. Measured ACLR of the three-stage WCDMA Doherty power amplifier with a device periphery of 1:2:4 versus frequency at the maximum output power level (33 dBm at 1.95 GHz).

The measurement results of output power characteristics and linearity of the dynamic-biased three-stage Doherty power amplifier are shown in Figures 7.19 and 7.20. It is seen that the results have improved in all respects. The ACLR levels at the output power of 33 dBm have decreased from –35 dBc to –40 dBc, and from –47 dBc to –55 dBc, for ACLR1 and ACLR2, respectively. This is expected from the improved AM-AM characteristics of the dynamically biased amplifier compared to the fixed bias amplifier, as observed from the output power curves in Figure 7.19. It can be concluded that with a correct choice of device periphery and appropriate biasing, the performance of the multistage design can be as good as that of a perfect Class A/AB single-stage amplifier at maximum operating power, with improved PAE in the low-power region.

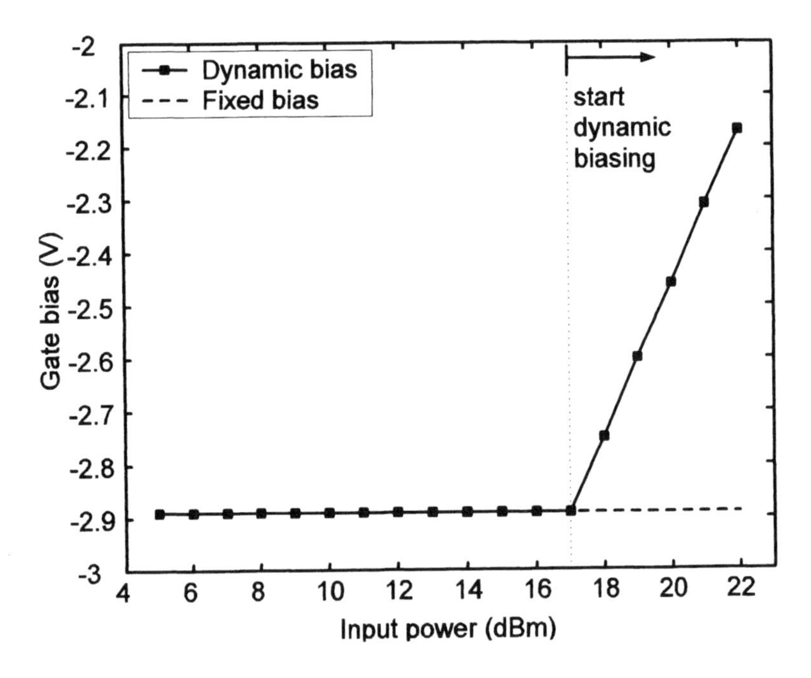

Figure 7.18. Bias voltage adjustment of peak amplifier 2 with increasing input drive level to alleviate AM-AM distortion.

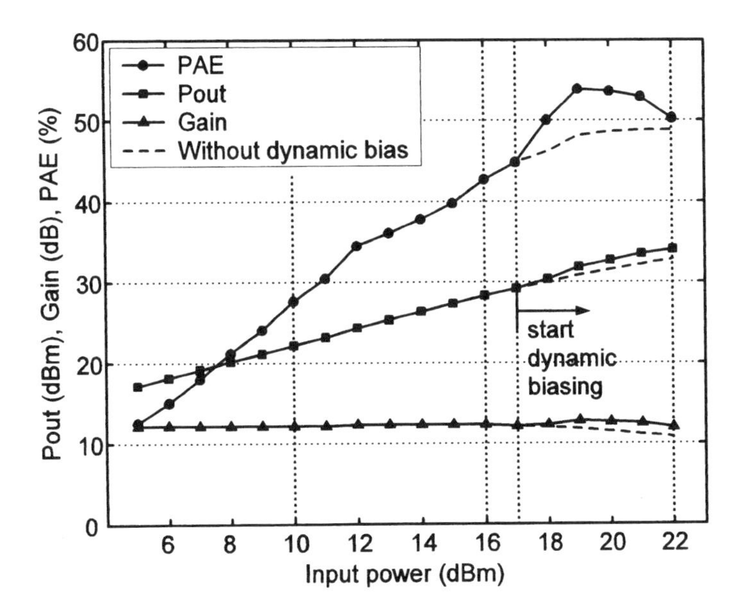

Figure 7.19. Measured output power, gain, and PAE of the three-stage WCDMA DPA with dynamic biasing applied to peak amplifier 2 (at 1.95 GHz).

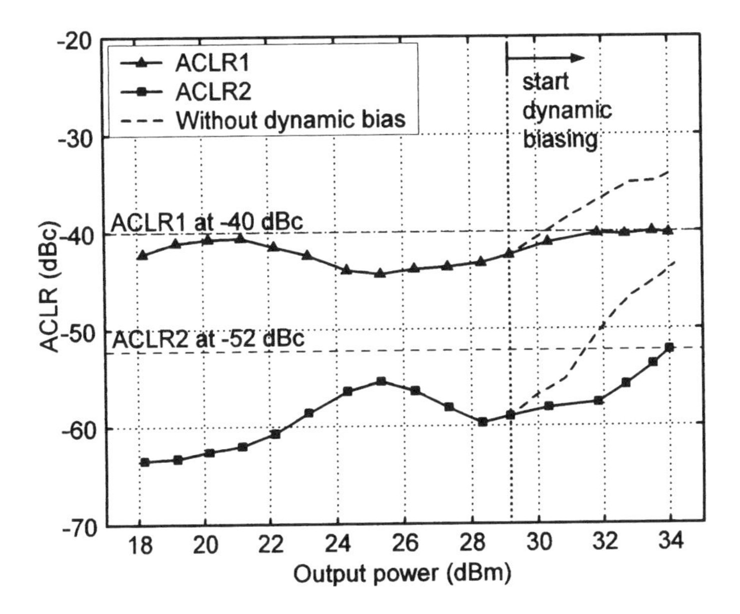

Figure 7.20. Measured ACLR of the three-stage WCDMA DPA with dynamic biasing applied to peak amplifier 2 (at 1.95 GHz).

REFERENCES

1. L. R. Kahn, Single sideband transmission by envelope elimination and restoration, *Proc. IRE*, **40**: 803–806 (July 1952).
2. G. Hanington, P. F. Chen, and P. M. Asbeck, Microwave power amplifier efficiency improvement with a 10 MHz HBT DC-DC converter, *IEEE MTT-S Int. Microwave Symp. Digest*, 1998, pp. 589–592.
3. G. Hanington, P.-F. Chen, P. M. Asbeck, and L. E. Larson, High-efficiency power amplifier using dynamic power-supply voltage for CDMA applications, *IEEE Trans. Microwave Theory Tech.*, **47**: 1471–1476 (Aug. 1999).
4. W. H. Doherty, A new high efficiency power amplifier for modulated waves, *Proc. IRE*, **24**: 1163–1182 (Sept. 1936).
5. H. Chireix, High power outphasing modulation, *Proc. IRE*, **23**(11): 1370–1392, (1935).
6. S. C. Cripps, *RF Power Amplifiers for Wireless Communications*, Artech House, Norwood, MA; 1999.
7. F. H. Raab, Efficiency of outphasing RF power amplifier systems, *IEEE Trans. Commun.*, **33**: 1094–1099 (Oct. 1985).
8. X. Zhang, L. E. Larson, and P. M. Asbeck, *Design of Linear RF Out-phasing Power Amplifiers*, Artech House, Norwood, MA; 2003.
9. F. H. Raab, Efficiency of Doherty RF power-amplifier systems, *IEEE Trans. Broadcast.*, **BC-33**: 77–83 (Sept. 1987).
10. M. Iwamoto, A. Williams, P.-F. Chen, A. Metzger, L. Larson, and P. Asbeck, An extended Doherty amplifier with high efficiency over a wide power range, *IEEE Trans. Microwave Theory Tech.*, **49**: 2472–2479 (Dec. 2001).

INDEX